D0340237

AUTOMATE THIS

AUTOMATE

HOW ALGORITHMS CAME TO RULE OUR WORLD

CHRISTOPHER STEINER

PORTFOLIO / PENGUIN

For Joseph Henryk Scott

PORTFOLIO / PENGUIN
Published by the Penguin Group
Penguin Group (USA) Inc., 375 Hudson Street,
New York, New York 10014, U.S.A.
Penguin Group (Canada), 90 Eglinton Avenue East, Suite 700,
Toronto, Ontario, Canada M4P 2Y3
(a division of Pearson Penguin Canada Inc.)
Penguin Books Ltd, 80 Strand, London WC2R 0RL, England
Penguin Ireland, 25 St. Stephen's Green, Dublin 2, Ireland
(a division of Penguin Books Ltd)
Penguin Books Australia Ltd, 250 Camberwell Road, Camberwell,
Victoria 3124, Australia
(a division of Pearson Australia Group Pty Ltd)
Penguin Books India Pvt Ltd, 11 Community Centre, Panchsheel Park,
New Delhi–110 017, India
Penguin Group (NZ), 67 Apollo Drive, Rosedale, Auckland 0632,
New Zealand (a division of Pearson New Zealand Ltd)
Penguin Books (South Africa) (Pty) Ltd, 24 Sturdee Avenue,
Rosebank, Johannesburg 2196, South Africa

Penguin Books Ltd, Registered Offices:
80 Strand, London WC2R 0RL, England

First published in 2012 by Portfolio / Penguin,
a member of Penguin Group (USA) Inc.

1 3 5 7 9 10 8 6 4 2

"Cannot the fervor," "Water snakes writhe," and "The fire crackles loud" from *Comes the Fiery Night: 2,000 Haiku by Man and Machine* by D. H. Cope (CreateSpace, 2011). Used by permission of the author.

Library of Congress Cataloging-in-Publication Data

Steiner, Christopher.
Automate this : how algorithms came to rule our world / Christopher Steiner.
p. cm.
Includes bibliographical references and index.
ISBN 978-1-59184-492-1
1. Finance—Computer simulation. 2. Finance—Decision making—Data processing. 3. Capital markets—Computer simulation. 4. Algorithms. 5. Computer simulation. I. Title.
HG106.S74 2012
332.01'13—dc23
2012018021

Printed in the United States of America
Set in Minion Pro
Designed by Pauline Neuwirth

CONTENTS

INTRODUCTION

IN EARLY APRIL 2011, MICHAEL EISEN, an evolutionary biologist at the University of California at Berkeley, logged on to Amazon .com to buy an extra book for his lab. He was after *The Making of a Fly,* by Peter Lawrence, about the genetic development of a fly from a single-celled egg to a buzzing, flying insect. The 1992 book, though out of print, remains popular with academics and graduate students. Eisen was used to paying $35 to $40 for a used copy. But on this day, April 8, there were two established Amazon sellers offering new, unused copies of the book for quite a bit more than he wanted to spend: $1,730,045 and $2,198,177.

Eisen assumed the price was a mistake or a joke; nobody, not even the author, he speculated, would put such a value on the book. He checked back on Amazon the next day to find that, rather than returning to normal, the prices for the book had risen to $2,194,443 and $2,788,233. On the third day, the prices ascended to $2,783,493 and $3,536,675. The escalation continued for two weeks, with the price peaking on April 18 at $23,698,655.93. And a buyer still had to foot a

$3.99 shipping bill. The next day, on April 19, the prices for the book fell, settling at $106.

But why had Amazon been selling an arcane book on fly genetics for nearly $24 million? Had it suddenly become hot with billionaire collectors? Did it contain clues to find treasure? Had it become the 1869 Château Lafite Rothschild of books? What had actually happened, in fact, was that the unsupervised algorithms that priced books for the sellers, both of whom sold thousands of titles on Amazon, got into something of a price war. One of the sellers' algorithms had been programmed to price the book slightly higher than the competitor's price. The second algorithm, in turn, always increased its price to be closer to that of the higher seller, prompting the first algorithm to respond with another price hike. And on it went until the books were as expensive as Manhattan penthouses. Things didn't return to normal until a human being stepped in and overrode the system. If this tale were an outlier, it would be a harmless anecdote about algorithms gone bad. Almost a year earlier, though, something far stranger had taken place.

Early in the day on May 6, 2010, the world's stock markets were aggravated lower by unsettling developments in Greece, where rioters protesting government spending cuts had engulfed Athens. Many feared that Greece would default on its debts, which would legitimatize fears of a global depression. By midmorning in New York, the American stock markets had retreated 2.5 percent, already an exceedingly bad day.

But things soon swerved from bad to inexplicable.

At 2:42 p.m. on the East Coast, the markets began to shudder before dropping into a free fall. By 2:47 p.m.—a mere three hundred seconds later—the Dow was down 998.5 points, easily the largest single-day drop in history. Screens tracking the Dow Jones Industrial Average, the most followed stock index in the world, looked like they'd been hacked by a practical joker. Nearly $1 trillion of wealth fell into the electronic ether.

CNBC's coverage of this five-minute span started with anchors and guests breezily chatting about riots in Greece. The Dow at that moment was down 200 points, as it had been most of the day. When anchor Erin Burnett brought on star commentator Jim Cramer, the Dow sat 350 points down. When it went 500 points down, Cramer, who is an animated man if he's anything, calmly muttered, "Things are getting interesting."

Within three minutes, viewers heard Burnett's voice breaking into Greece coverage to say the Dow was now down 800 points. Burnett, with disbelief flecking her words, paged CNBC reporter Scott Wapner on the floor of the New York Stock Exchange: "Scott, what are people saying? Now you're down 800."

"When I asked them what the heck was going on down here . . . I don't know," Wapner replied, stumbling to find his words. "There is fear. This is capitulation. This is classic capitulation."

CNBC then went back to Cramer, who remained oddly serene, even though the Dow, in a span of three minutes, had fallen nearly 1,000 points. "I just sat down," Cramer said, referring to his arrival on the set, "and that was five hundred points ago."

Burnett, the anchor, then brought up Procter & Gamble, whose blue-chip stock, in a span of minutes, dove 25 percent from $62 to $47. This brought an end to Cramer's stoicism.

"It can't be there. That is not a real price," Cramer proclaimed exasperatedly. Then, indignantly, he barked, "Okay, just go buy Procter." He wheeled to look directly at the camera, imploring viewers, "Go buy it right now. Just go buy it!"

The pros heeded that exact advice. Just as Cramer sensed that stocks had quickly gotten way too cheap, so did others. After the Dow hit its nadir, down 998.5 points, it began racing back up just as swiftly as it had fallen. The market rose faster, in fact, than it had gone down. Within a minute, the Dow, which rarely moves 300 points in a whole day, shot up that exact amount. The mere thought of it confounded everybody watching.

At Magnetar Capital, a large hedge fund in Evanston, Illinois, that manages $7 billion, senior executives rushed out of their offices to the firm's trading floor after hearing a commotion. "Everybody down here just started yelling, 'Buy, buy, buy,'" says a Magnetar trader. "It didn't matter what you were trading—just buy."

On the East Coast, at First New York Securities, the trading floor echoed that of Magnetar's. More than 120 traders jockeyed to get orders filled, yelling one simple word: "Buy."

"My first reaction was, 'Okay, so this is an error, how do I take advantage of it?'" says Tom Donino, head of trading at First New York. "I've been doing this for twenty-five years and I've never seen anything like that in my life."

Some share prices crashed to one penny—as in $0.01—rendering billion-dollar companies worthless, only to bounce back to $30 or $40 in a few seconds. Other stocks swung wildly up. At one point, Apple traded at $100,000 a share (up from about $250). The market had been gripped with violent turbulence and nobody knew why. Whatever the problem, it couldn't all be dumped on one giant erroneous order or a single rogue trader. The action came so furiously that some traders and market watchers, perhaps on a bathroom break or lost in a cup of coffee, missed it completely.

Back at CNBC's studios, Burnett and Cramer watched as the Dow recovered 500 points in less than three minutes. Cramer remained strangely unimpressed by the spectacle, although such violent swings had never before taken place on the stock market.

"The machine obviously broke, the system obviously broke," Cramer said, a touch of disgust in his voice.

Burnett's journalistic fiber had her more excited. "But the fact that after all this, that that could have just happened, is an absolutely stupendous story," she exclaimed.

"I think it's a great story," Cramer said flatly. "It's the greatest story never told. You'll never know what happened here."

Cramer wasn't wrong. As of this writing, there is still no consensus on the exact root of what became known as the Flash Crash. Some of the blame was directed at a Kansas City money manager whose algorithm sold off $4 billion worth of stock futures too quickly, sparking other algorithms to do the same. Some blame an unknown group of traders who conspired to send things down all at once through the use of coordinated algorithms. Some believe it was simply an old-fashioned panic, not unlike what the world witnessed in 1929. What's for certain, however, is that the market couldn't have moved so far so fast if algorithms, which act independently of humans and require less than a second to place and complete a trade, didn't own the market. But they do.

Algorithms normally behave as they're designed, quietly trading stocks or, in the case of Amazon, pricing books according to supply and demand. But left unsupervised, algorithms can and will do strange things. As we put more and more of our world under the control of algorithms, we can lose track of who—or what—is pulling the strings. This is a fact that had sneaked up on the world until the Flash Crash shook us awake.

Algorithms entered evening newscasts through the door of the Flash Crash, but they didn't leave. They soon showed up in stories about dating, shopping, entertainment, medicine—everything imaginable. The Flash Crash had merely been an augur for a bigger trend: algorithms are taking over everything.

When a process on the Web or inside a machine happens automatically, a pithy explanation often comes with it: "It's an algorithm." The classical definition of an algorithm says the device is a list of instructions that leads its user to a particular answer or output based on the information at hand.

One could, for instance, write an algorithm for determining what jacket to wear to work in the morning. Inputs: temperature, presence of rain, presence of snow, wind speed, distance and pace you plan to walk, sun or cloud cover. An input of 25 degrees, light snow, 20 mph wind, cloud cover, and a short walk of two blocks might produce an output of, say, your down-filled Gore-Tex parka. That's likely the coat you would have plucked from the closet on your own, but the invasion of algorithms starts with simple tasks. These algorithms operate much like decision trees, wherein the resolution to a complex problem, requiring consideration of a large set of variables, can be broken down to a long string of binary choices. Each piece of required data pushes the process to another choice, or node, and closer to producing an output.

This rudimentary definition of algorithms, however, gives little justice to the colossal webs they have become thanks to computers. In this book, I often refer to multiple linked algorithms all aimed at performing one task as bots. These bots feature thousands of inputs, factors, and functions. The most complicated among them are like neurons firing in your brain: they spin up and spin down based on need, they're dynamic, and they're capable of self-improvement.

Math makes possible all of these algorithms that have come to invade and almost run our lives. For centuries, math was something we drew on in making observations about our world. Now it is a potent tool we use to shape our planet, our lives, and even our culture.

As algorithms and the math behind them became the standard on Wall Street, other less-affected fields drew the attention of mathematicians, engineers, and physicists—a group Wall Street came to call quants (after quantitative analyst). These quants and programmers now scout new industries for soft spots where algorithms might render old paradigms extinct, and in the process make mountains of money.

The bounds of algorithms get pushed further each day. They've displaced humans in a growing number of industries, something they often do well. They're faster than us, they're cheaper than us, and, when things work as they should, they make far fewer mistakes than we do. But as algorithms acquire power and independence, there can be unexpected consequences. Perhaps Pandora, the Internet radio outlet that uses algorithms to learn users' musical tastes, is aptly named. According to Greek mythology, Pandora, the first woman on earth, was given a beautiful box that Zeus warned her not to open. She opened it anyway, releasing evil across the planet. Only one thing remained inside: hope.

While there are many reasons to worry about the impending age of algorithms, there are also, as Pandora instructs us, plenty of reasons for hope. Algorithms will enable myriad improvements to our world, like better radio, of course, along with better customer service calls, better Friday dinner dates, better CIA intelligence, and better ways to sniff out cancer before it can kill.

Some algorithms' roots trace to the field of artificial intelligence. They may not be intelligent and self-aware like Hal 9000 (*H*euristically programmed *AL*gorithmic computer), the machine from the movie *2001: A Space Odyssey* (1968), but algorithms can evolve. They observe, experiment, and learn—all independently of their human creators. Using advanced computer science techniques such as machine learning and neural networking, algorithms can even create new and improved algorithms based on observed results. Algorithms have already written symphonies as moving as those composed by Beethoven, picked through legalese with the deftness of a senior law partner, diagnosed patients with more accuracy than a doctor, written news articles with the smooth hand of a seasoned reporter, and driven vehicles on urban highways with far better control than a human.

And what will become of our duties as humans, our employment? Algorithms will have a say here too. Jobs we once blamed other countries for stealing are now being snatched away by faceless piles of computer code. It's no coincidence that the most upwardly mobile people in society right now are those who can manipulate code to create algorithms that can sprint through oceans of data, recognize millions of faces, and perform tasks that just a few years ago seemed unthinkable.

HACKERS: THE NEW EMPIRE BUILDERS

There seem to be two divergent definitions of the term *hacker* floating about the modern lexicon. To some, hacking has come to mean something inherently criminal—a programmer traversing electronic property meant to be off-limits. Depictions of hackers from the 1980s portrayed fast-typing miscreants breaking into secret government data hoards, banks, and all sorts of places they didn't belong.

Within the modern tech world, however, the word carries good connotations rather than bad. This is the definition I use in this book. A hacker in the parlance of Silicon Valley is somebody who writes computer code with a deft hand, easily translating concepts from the whiteboard into algorithms that can make hard decisions, trade stocks, drive cars, sort college applicants, or play poker with the best humans in the world.

Algorithms are at the center of our story, but it's hackers' code that makes them possible; it's what gives them life and the ability to affect millions of people. The code behind the algorithm can be thousands of lines long, linking mathematical functions to user inputs such as yes/no decisions or the real-time mining of data. To hack something of moderate difficulty—a standard Web site, say, or an automated e-mail program—is a skill possessed by millions of people across the world. To hack something innovative, to conceive and write an elegant algorithm that solves human problems, takes special talent.

I came to this story of algorithmic revolution in a roundabout way. I haven't written meaningful code since 2002 when I composed simple algorithms in C, a computer language, during my last months working as an engineer. In 2003, I followed an itch to write and went into journalism. My first job was with the *Chicago Tribune,* where as a newsroom grunt I chased all manner of things from fires to murders to business stories. Serendipity landed me a job in the Chicago bureau of *Forbes* magazine. Tech became my natural focus and I spent much of my time sniffing about the country for new companies and paradigms to write about.

Spending so much time with young entrepreneurs and engineers can be invigorating. I met a lot of people who left stable jobs or career paths to try to build companies and products all their own. At some point, I became affected. For years, I threw around startup ideas. In the fall of 2010, I conjured something with my friend Riley Scott that seemed like it could work. The idea eventually became Aisle50, a Web site that offers deals on grocery items for consumers to purchase before they head to the store. While building Aisle50, we applied and were accepted to be part of Silicon Valley's Y Combinator, which helps start-ups get off the ground with funding, mentorship, and connections to investors.

A good percentage of my cohort at Y Combinator picked up programming by the time they were fourteen years old. When they reached college, often an elite university, they could already string together thousands of lines of code, create the guts of a stable application, or bolt up an original Web site design in a matter of hours. During the three months I spent there in the summer of 2011, I learned how Mark Zuckerbergs are created: not from spontaneous explosions of intellect and technology, but from years and years of staring at a computer screen, getting to know code as intuitively as a seasoned copy editor knows idioms, punctuation, and style.

These stylists of code and writers of algorithms are the preeminent entrepreneurs of this generation. The builders of new empires no longer come from business school—they come from engineering and computer science labs, where long nights of staring at coding assignments result in the hacking skills required to build innovative algorithms and the companies they can propel. In some cases, these algorithm builders leave college early or don't bother to go at all; they learned enough when they were fifteen, sixteen, and seventeen to know how to build what they envision.

There now exist all sorts of paths for people to learn programming at young ages. Countless Web communities and chat rooms are dedicated solely to the writing of code and the construction of algorithms. One company from my Y Combinator group, Codecademy, hit upon an idea so popular—a well-designed site for learning how to program online—that it drew more than 200,000 users in its first two weeks after launching. Just six months after launch, Codecademy landed a partnership with the White House to promote computer programming. This is the new reality we live in: a twenty-one-year-old with the tools to conceive complex algorithms can form a partnership with the president of the United States.

Y Combinator is just a microcosm of a movement, often pushed forward by youth, that is putting algorithms inside everything we do. During the last couple of decades, this class of revolutionaries has led the way in fashioning algorithms that solve problems, make money, and, in sweeping fashion, take jobs away from people. This book is about how we got here and how things will continue to unfold.

The resources available to aspiring bot builders today, even compared with those from just five years ago, are astounding. Building innovative algorithms isn't easy. But it gets less difficult every day. For this reason, the people with the ability to create these bots—the best math, science, and hacker minds—look out upon a promising future. The apocryphal story of Bill Gates finding computer time by finagling his way into labs at the University of Washington during odd hours is something that won't be repeated because it won't be necessary. Access to technology now comes easily.

When I started working on this book in late 2009, my focus and my reporting revolved solely around algorithms on Wall Street. Then I had lunch with Bill Bennett, a friend and a professor at Northwestern's Kellogg School of Management. I told him I was writing about how algorithms had taken over our stock markets. Bill thought that was fine. But as he looked past his plate of cashew chicken and across the street at the building that houses Magnetar, the hedge fund, he said something that would steer this book. "The real story," he said, "is how they will take over everything."

Bill was right. This story only begins on Wall Street. From there, it goes everywhere.

1

WALL STREET, THE FIRST DOMINO

ON A DAY IN EARLY 1987, a man who worked for the Nasdaq stock market—let's call him Jones—showed up in the lobby of the World Trade Center. He found the appropriate elevator bank for his floor and pressed the up button. He was making a routine visit to one of the Nasdaq's fastest-growing customers. Jones knew what to expect. One Wall Street stock-swapping crew is the same as the next: a small lake of white male faces backed by Ivy League educations and appetites for profit. Nothing to see, really.

As he walked down the hallway toward the office suite's front door, he braced himself for the testosterone and chaos that awaited him. Trading floors aren't anomalous from game shows that stuff a person in a glass box full of blowing money—except that in the case of a trading floor some of the fluttering bills are losers. The deft players can discern winning bills from losing ones while maintaining a frenzied pace.

A receptionist greeted him and retreated to another room to fetch his host. When she returned, a short, dapper man with a full head of silvering hair accompanied her. Thomas Peterffy's blues eyes warmly greeted Jones. He spoke with an accent.

Jones couldn't have known that Peterffy would later become a man worth more than $5 billion, one of the richest people in America. He was still at that point a Wall Street upstart. But his trading volume had been streaming upward, and so had his profits. Jones was always curious as to how people like Peterffy figured out ways to beat the market so consistently. Had he hired the sharpest people? Did he have a better research department? Was he taking giant risks and getting lucky?

What Jones didn't know was that Peterffy wasn't a trader at all. He was a computer programmer. He didn't make trades by measuring the feelings of faces in the pit, the momentum of the market, or where he thought economic trends were leading stocks. He wrote code, thousands of lines of computer language—Fortran, C, and Lisp—all of it building algorithms that made Peterffy's trading operation one of the best on the Street, albeit still small. He was chief among a new breed on Wall Street.

As Peterffy led the way onto his trading floor, Jones grew confused. The more he saw—and there wasn't much to see—the more flummoxed he became. He had expected a room bursting with commotion: phones ringing, printers cranking, and traders shouting to one another as they entered buy and sell orders into their Nasdaq terminals. But Jones saw none of this. In fact, he saw only one Nasdaq terminal. He knew the volumes that Peterffy did—and they were big. How was this possible? Who was making all those trades?

"Where is the rest of the operation?" Jones demanded. "Where are your traders?"

"This is it, it's all right here," Peterffy said, pointing at an IBM computer squatting next to the sole Nasdaq terminal in the room. "We do it all from this." A tangle of wires ran between the Nasdaq machine and the IBM, which hosted code that dictated what, when, and how much to trade. The Nasdaq employee didn't realize it, but he had walked in on the first fully automated algorithmic trading system in the world.

Peterffy's setup didn't merely suggest what to trade, as other systems had in the past. It didn't simply pump out trades that humans would later carry out. The computer, by way of a surreptitious hack into the trading terminal, made all of the decisions and executed all of the trades. No humans necessary. Its trading partners, though, were 100 percent human—and they were getting drubbed.

With the hacked data feed coming from the Nasdaq terminal, Peterffy's code was able to survey the market and issue bids and asks that could easily capture the difference between the prevailing price at which buyers would buy and sellers would sell. That difference, called the spread, could grow past 25 cents a share on some Nasdaq stocks at that time, so executing a pair of 1,000-share orders—one to buy at $19.75 and one to sell at $20.00—resulted in a near-riskless $250 profit.

The trades were even less risky for Peterffy because he utilized machines to execute them. One of the main dangers at that time to market makers, who constantly maintain offers to buy or sell a stock, was leaving stale quotes up after a market shift. Most market makers could only be as fast as their traders, who had to read new prices on a computer screen, assimilate that information, make a decision on how to change their prices, and then cancel orders and type new prices into their Nasdaq keyboards. A trader could fall a few steps behind the market just by taking too many bites of a tuna sandwich or having a chuckle with a colleague. Peterffy's computer didn't need a lunch. It stepped in tandem with the market's zigs and zags, mitigating a high percentage of his market risk, something a human simply couldn't do.

Peterffy's operation marked a new dawn on Wall Street, as programmers, engineers, and mathematicians mounted a two-decade invasion in which algorithms and automation, sometimes incredibly complex and almost intelligent, would supplant humans as the dominant force in our financial markets.

Jones stood agape. Where Peterffy saw innovation, Jones saw somebody breaking the rules with a jury-rigged terminal.

"You can't do this," Jones said.

The Nasdaq had no trading floor; all of its trades took place over the phone or on its computer network that took users' orders as they entered them on the keyboard of a dedicated Nasdaq terminal. Peterffy had taken the incoming data wire meant for the terminal and spliced it, soldering the split end into a circuit board that his team of programmers and physicists had built from scratch and embedded into the motherboard of an IBM PC. The IBM ran software that Peterffy wrote himself. As the PC got data from the Nasdaq wire splice, its algorithms analyzed the market and made quick trading decisions, firing these trades back through a tangle of wires that wound their way into the innards of the Nasdaq terminal. Peterffy, unbeknownst to anybody until that moment, had hacked the Nasdaq.

The Nasdaq didn't need word of this contraption, this mad scientist's laboratory, reaching the market. Would other traders be comfortable knowing they were matching wits with algorithms powered by an IBM rather than other gut-following gamblers? The Nasdaq didn't want to find out.

"The terminal needs to be disconnected from this IBM and all orders need to go through the keyboard, typed one by one—just like the rest of our customers," Jones said.

Jones left. Peterffy stood in his office contemplating what might be the end of his business. The Nasdaq had given him a week to comply with the inspector's edict. The thought of dismantling his machine wrenched Peterffy. He had little interest in going out and finding traders, even young and cheap ones, to sit in chairs and punch orders into Nasdaq terminals. It had taken him years to wring the human element and its capricious whims out of his operation. It would be difficult to reinject people, their errors, their laziness, and, most important in this case, their slow typing back into the process and expect the same results. His operation was going to lose most of its efficiency overnight. There had to be a better option.

Before he went to sleep that night in his Upper East Side apartment, a solution came together in his head. It wouldn't be easy, but it offered possibilities. Peterffy thought he could pull information from the Nasdaq terminal without touching the machine. No spliced wires, no attached circuit boards, none of that. But how to do it? He asked his crew of engineers and physicists if they could build something that read data straight off the screen, like a camera, and then translated that information into electronic bits and sent it to the waiting IBM PC. The answer was yes.

But the data feed was only half the problem. How would Peterffy execute his trades without having a team of people sit at Nasdaq terminals? He could not send a wire back into the machines as he had done before. No, the trades had to go through the keyboard, just as the Nasdaq had ordered. Peterffy had an idea, a crazy idea. But could such a thing work?

During a frantic week, Peterffy and his best engineers welded metal, wrote code, and soldered wires. They affixed a large Fresnel lens to the face of the Nasdaq terminal to enlarge the screen's text. A camera was placed a foot from the lens. From the camera, a wire led to a computer sitting adjacent to the apparatus. In just a few days, Peterffy and his programmers wrote software that would decode the visual data streaming in from the camera. From there, the data could be plugged into Peterffy's existing algorithms that once used the direct wire from the Nasdaq terminal.

A new wire now came out of the IBM and, instead of slithering into the case of the Nasdaq terminal, ran into a nest of metal rods, pistons, and levers hovering above the terminal's keyboard. If the camera and screen-reading rig appeared odd, this piece of the system was downright bizarre. It evoked the intricate mechanical looms of the Industrial Revolution. The device was an automated typing machine, built from scratch. As orders came from the computer, the machine's rods rapped the terminal's keys in staccato bursts. Orders flew out one after another, with dozens logged in under thirty seconds.

The Nasdaq said trades had to be typed, but they didn't specify who had to type them. Peterffy's team had created a trading—and typing—cyborg. And it had taken them six days. He had obeyed the letter of the law, but had, by any definition, violated its spirit. But that didn't worry him much. Wall Street is nothing if not a den of loopholes, work-arounds, and back doors that favor the most inventive of finaglers.

The Nasdaq inspector returned a week later, as promised. Peterffy met him at the elevator and led him down the hallway to his trading room. A steady clinking emanated from the door. Where once had been silence, there now arose a racket, just as it should be on a trading floor. Peterffy led Jones through the door and, with a flourish, gestured to his creation. The Nasdaq man drank in a scene that could have been conjured by Jules Verne.

"What is this?" the man said.

Peterffy explained that his machine put in the trades just like the Nasdaq requested—by keyboard input, one at a time. Just at that moment, the market came to life, as did the machine. Peterffy's program began trading so rapidly that the typing apparatus fired like a machine gun. The continuous stream of orders caused the machine to mash the keyboard furiously, creating a chaotic din that made conversation impossible. Each time it seemed a lull had set in, the machine would fire out more orders, refilling the fleeting silence. The whole operation was yet another marvelous rule bypass brought to the world by the flexible minds on Wall Street.

"He did not like this one bit," Peterffy recalls.

Feeling he had nothing to lose at this point, Peterffy offered to build a mannequin into the contraption and make the doll strike the keys. He was half joking, but he would have done it. Jones's face remained tight.

As the man shook his head, Peterffy grimaced. He had built the fastest trading machine in the world and he now fully expected to have to take it apart. After stewing for several minutes, the Nasdaq man stalked out of Peterffy's offices wordlessly. Peterffy prepared for the worst: a fiat from the Nasdaq that would ban his technology from the market. But Jones never came back, and the phone call Peterffy feared never came. His operation left intact, Peterffy, who had begun with less than $100,000 several years before, made $25 million in 1987.

Peterffy was still a Wall Street minnow in 1987, but he was one of the lead fish in a new pack of market players who were as adept writing complex code, soldering semiconductor chips, and employing math as they were negotiating labyrinthine market structures. What Peterffy had done was simple in theory and complicated in execution: he had taken the brains of the smartest traders and found a way to express those smarts in a series of algorithms. His programming included all the elements that a crack human trader weighed in making a decision. But the computer took far less time to do the math, check the prices, and pull the trigger.

There would be others to wield software, code, and swift computers to beat the market, but Peterffy's innovations, from keyboard-hammering pistons to hacked data feeds, sparked this revolution. Today, 60 percent of all trades are executed by computers with little or no real-time oversight from humans. The story Peterffy weaved is unique on Wall Street. He wasn't a big-shot trader who presciently hired programmers to further his dominance. Nor was he a Wall Street player who taught himself coding to gain an advantage over the rest of the field. What made Peterffy unique was that he was a programmer—and a good one—long before he ever understood how a stock option worked or how shares of different companies often move in tandem.

Using his programming skills, his understanding of math, and his mastery of writing complex code, Peterffy created layered algorithms to disrupt a field new to him: the trading floors of Wall Street. This disruptive hacker paradigm has played out across varying parts of our world during the latter years of the twentieth century and opening of the twenty-first: a deft engineer of computer code and algorithms takes an interest in a new field, develops expertise, and, by applying computer science and clips of code that mimic their human forerunners, topples industries, companies, standards, and the old guard. The ability to create algorithms that imitate, better, and eventually replace humans is the paramount skill of the next one hundred years. As the people who can do this multiply, jobs will disappear, lives will change, and industries will be reborn. It's already happened, and it will continue. And as with any trend, this one follows the money. That's why it began on Wall Street, in no small part thanks to a Hungarian immigrant.

THE MAKING OF A HACKER, THE HARD WAY

Thomas Peterffy may have become one of the most important figures in this thirty-year tale of creeping algorithmic takeovers, but he was born not into a world of access but into the sounds of the bloodiest war in human history. Delivered in the basement of a Budapest hospital during a 1944 bombing raid, Peterffy was raised by his mother and grandparents after his father fled the Soviet-backed communists who took over Hungary following World War II. He remembers a childhood of fear, of relatives disappearing, of moves in the middle of the night, and of the constant threat of starvation, freezing, or worse.

Peterffy discovered capitalism in high school, when he sold smuggled sticks of Juicy Fruit gum to students at a 500 percent markup. At thirteen, he organized several crews of boys who would raid abandoned buildings and rubble heaps for any metal that could be scavenged. He bartered with men three or four times his age and found the best prices for his scrap, splitting the proceeds among his crews. He later turned his industrious focus to stamp buying, selling, and trading. The irregular market, in which some stamps sold for more in one place than they did in another, enthralled him. He had discovered arbitrage, where one takes advantage of similar asset markets with disparate prices. An arbitrageur buys where the price is lower and sells where the price is higher—a strategy that, in far faster form, comprises the backbone of many modern high-speed trading operations.

After graduating high school, Peterffy studied advanced geometry in a technical school for surveyors, with a long-term goal of college and a degree in civil engineering. But his education was derailed in 1965 when, at twenty-one, he got a short-term visa to visit some distant relatives in West Germany. He seized this opportunity in Americanized Germany to apply for a U.S. immigration permit, which he eventually received. Boarding a plane to New York, the young Hungarian had little concept of Wall Street, let alone the computers and algorithms with which he'd conquer it.

He found lodging in a two-room apartment on the Upper East Side with a monk who had been excommunicated by the church for drinking and womanizing. The monk brought Peterffy into his business for a time as an unofficial notary public for Hungarian immigrants. Through the growing Hungarian immigrant network in New York City, Peterffy eventually found a new apartment and a job drafting plans at a civil engineering firm.

By the mid-1960s, computers had just begun to fall within reach of small businesses. The boxy machines soon flooded the offices of well-intentioned business owners who believed harnessing computing power would bring them more clients and more efficiency. But the grand arrival of computers in these places was met with a dearth of people who actually knew how to program them. Many computer buyers felt stymied when they discovered that, for all the fanfare, using a computer was hardly liberating. It was difficult. For that reason, many of the machines purchased during this era collected dust, unused and relegated to closets and back rooms.

When Peterffy's firm bought its first computer in 1964, nobody at the engineering company knew how to program it. Peterffy volunteered. He pored over the English manuals and began to pick up coding. He created simple algorithms executing the Pythagorean theorem, ones that determined angles using sine and cosine functions, and others that helped engineers prescribe the radii and slopes of joining roads. By 1966, he had built the firm a library of programs while collecting sixty-five dollars a week in salary. Picking up programming was easier for him than learning how to converse in English. He excelled.

Just as is the case now, good computer programmers in the late 1960s didn't want for jobs. Writing code offered Peterffy a clear path to more money and better work. His first opportunity came in 1967 when he left the engineering company for a position at Aranyi Associates, a firm that helped its Wall Street clients set up computer systems. The job, again, came through native networking; the firm's owner, Janos Aranyi, was a Hungarian immigrant.

On Wall Street, Peterffy confronted what we'll call phase zero of the algorithm story, which is to say that algorithmic bots had no place in this era of people's lives or on Wall Street. It was a world of human choices and human strategy executed by humans. A trader made a decision on the floor based on where he believed the market was headed, the securities he already had in his pocket, or, in many cases, an ambiguous gut feeling. He signaled his trade to another human, who took note of the transaction in a notebook. This was the world.

In his first finance job, Peterffy built algorithms allowing investors and traders to easily compare securities' different characteristics and values all at once. Bearing down with an intensity that came from being an immigrant with little else on which to focus, he made himself into a formidable programmer at a time when there existed few—and almost none of them on Wall Street.

After three years with Aranyi, Peterffy's computer skills landed him an interesting job offer with one of the best-known players in New York's markets: Dr. Henry Jarecki, a psychiatrist who had taken a fancy to trading commodities and during the 1960s built a large precious metals trading company that took the name of Mocatta Group, after Mocatta & Goldsmid, a three-hundred-year-old London firm with which Jarecki had originally partnered.[1] Jarecki started Peterffy with a $20,000 salary and a $4,000 bonus—about the equivalent of $145,000 in 2012 and good money for somebody a few years removed from sharing an unheated apartment with an excommunicated monk. Peterffy found the metals market intuitive and moved up quickly at Mocatta as his programming skill became indispensable.

THE ALGORITHM THAT CHANGED WALL STREET

With Jarecki in 1969, Peterffy introduced one of Wall Street's first so-called black boxes, which inhale market data, chew on it, then issue an instruction to their user, in this case whether to buy or sell. The calculus of an algorithm making trading decisions—really, any decisions—is built out of models, functions, and decision trees that at first are largely based on how humans make similar decisions. If Mocatta's traders usually dumped gold on Friday because they noticed that other traders got bullish before the weekend, then such preferences were written directly into the program. But it was also possible that Mocatta traders ignored their Friday selling rule when gold had been down the previous four days in a row. This condition too would be written into the computer program through a series of linked algorithms.

In 1969, Mocatta traders dialing on phones or shouting in pits made the actual trades, but the decision to buy or sell came straight from code Peterffy wrote. As good as Peterffy had become, there was one black box that had eluded him. Jarecki had asked him to come up with an algorithm that could peg exactly the correct price for options, whose prices, like anything else, obeyed no particular set of rules other than the vacillating whims of traders in the pit. Peterffy had spent months on the problem and produced little of use.

An option confers the right to buy or sell a given security at a set price before an expiration date. If IBM is trading at $100 and a trader is convinced it will shoot up during the next month, he might buy a call option that gives him the right to buy the stock at $100 during the next two months. The call option might only cost $10, so it's quite a bit cheaper than buying shares of IBM; a call option's downside is also limited—the most a trader might lose is simply the price of the option, or $10. The trader who sells the call option usually holds some IBM himself and is hedging against a dip by selling the call and pocketing that cash. He will have to surrender his shares for $100 if, say, the price shoots up to $110. Put options, conversely, give one the right to sell a stock at a set price in the future. So buying a put to sell IBM at $100 is a bearish bet, because it only succeeds if the price of IBM falls. Buying a call is bullish.

In the early 1970s, options had just begun to change hands with frequency. Mocatta was one of the firms leading the way. Jarecki and Peterffy had come to the conclusion that there were three main ingredients to pricing an option: the price that the option allowed one to buy or sell the stock for—called the strike price; the expiration date; and the volatility of the stock or metal. For instance, the options for a security whose price can swing wildly should be more expensive, because it's more likely that extreme strike prices will be hit. Other ingredients matter too, such as the prevailing risk-free interest rate, which forces the price of call options up and the price of put options down. Peterffy needed a way to express all of this in one elegant algorithm that rightly weighted each factor. It was a complicated math problem that he found nearly impossible to solve. He cycled through spurts of dejection and inspiration.

After working on the problem for more than a year, Peterffy devised an algorithm of differential equations that cleverly weighted all of the ingredients. He back-tested the algorithm to see if it would have made money in the past, but the data sets for commodities options at that point in history were limited. This was before computers handled such things adeptly and, more important, before the options market had much history. So Mocatta did the only thing it could: it started trading with the algorithm. It made money. The options markets weren't the giant realms they are today, so the algorithm wasn't able to harvest billions of dollars, but it gave Mocatta's traders a big edge. Most of the company's pit people didn't know where their trade orders came from; they knew only that their trades almost always worked. This is phase one of an algorithmic takeover: a computer, equipped with human-composed algorithms, analyzes inputs and issues marching orders to humans.

About a year after the men had put their algorithm to work, a thunderclap sounded above Wall Street. In 1973 Fischer Black and Myron Scholes, both professors at the University of Chicago, published a paper that included what would become known as the Black-Scholes formula, which told its users exactly how much an option was worth. Algorithms based on Black-Scholes would over the course of decades reshape Wall Street and bring a flock of like-minded men—mathematicians and engineers—to the front lines of the financial world. The Black-Scholes solution, quite similar to Peterffy's, earned Myron Scholes a Nobel Prize in 1997 (Black had died in 1995).

Change didn't happen overnight. The Black-Scholes formula, a partial differential equation, was brilliant. But most traders didn't peruse academic journals. Even if they did, employing the formula wasn't simple; it took significant math skills to wield. As few people as there were who understood Black-Scholes, even fewer knew that a Hungarian had written a similar algorithm that made money on almost every trade. Peterffy and Jarecki kept it quiet.

At a 2010 black-tie banquet in Chicago, a spry Jarecki spotted Scholes across the room nursing a cocktail and lightly conversing. Jarecki made his way over to the Nobel laureate. "You know, you still have our Nobel Prize," Jarecki said to Scholes. The remark elicited a dry grimace. "He was not amused," Jarecki says.

For traders who understood it, Black-Scholes gave them a way to calculate the exact price at which options should be traded. It was like having a cheat sheet for the market. There was money to be made by anybody who could accurately calculate each factor within the Black-Scholes formula and apply it to options prices in real time. Traders using the formula would sell options that were priced higher than the formula stipulated and buy ones that were priced lower than their fair price. Do this enough times with enough securities and a healthy profit was virtually guaranteed.

TO BE A WALL STREET HACKER IN 1980: PERFECT PLACE, PERFECT TIME

The late 1970s marked the faint dawn of the hacker era on Wall Street, when algorithms began to step in front of humans, a trend that has come to dominate all financial markets in every corner of the world. Wall Street began to peel away an increasing share of the best math and science minds in the United States and put them to work programming and conceiving trading algorithms. Before Black-Scholes got hold of the market, there had always been a slow trickle of engineers and mathematicians into lower Manhattan, but they were rare.

The engineering and science halls at MIT, Harvard, and other elite universities became places where recruiters wrestled for leverage. Wall Street would always be there, around the corner, skulking about with its promises of cash, glamour, and bonuses. The financial industry would eventually succeed not only in pulling away promising young talent from research universities and tech companies but also in luring accomplished and veteran engineers and scientists from cushy positions of prominence in tech and academia.

Why did Wall Street want all of this talent? For speed. Those who are first in to a good trade win—and algorithms running on computers will always beat a human looking for the same kinds of trades. Being able to express algorithms in the form of code in the late 1970s was not a common skill. This is why Peterffy, an outlier with his fifteen years of coding experience and ten years of plying the markets, found himself in as perfect a position as anybody in the world to lead the coming changes on Wall Street.

Having written the options algorithm, Peterffy began to build his programming department, bringing on more coders as their utility expanded in the market. Mocatta, through Jarecki's trade smarts and Peterffy's mastery of algorithms, began to make millions of dollars and became one of the most powerful commodities traders in the world. As

Mocatta grew, so did Peterffy's hacker army. By 1975, they employed fifty programmers, making the company one of the few mainstays of programming knowledge on Wall Street.

What were all of these programmers doing? Most of them were helping Peterffy put trading algorithms, conceived by himself and Jarecki, into code that could be executed by a computer. As the algorithms grew more complicated, better code and more programmers were needed.

Algorithms can start as simple things. Take one for, say, doing the laundry. The inputs could be as simple as (1) the weight of the laundry and (2) the type of fabric. The algorithm would take the weight input and, if it were below one pound, set the washing machine's water level to low. It would then take the fabric input of "cotton" and set the machine's water temperature to hot for the wash and cold for the rinse. Different weights and fabrics would elicit different washing programs from the algorithm. But what if the algorithm had to also pick the exact right wash according to the laundry's colors, stains, presoak needs, drying time, and detergent types? Composing such an algorithm is well within the reach of any sophomore-year computer science student, but is still multitudes more difficult and layered than the first algorithm whose only inputs are weight and fabric. Humans can easily dispatch a load of laundry with multiple washes and needs without stretching their brains; writing a computer algorithm with the same capabilities, though, takes skill. All of a sudden, there are hundreds of possible solutions based on inputs. Each one has to be accounted for and rightly categorized for the algorithm to work correctly.

In this way, algorithms can be looked at as giant decision trees composed of one binary decision after another. Almost everything we do, from driving a car to trading a stock to picking a spouse, can be broken down to a string of binary decisions based on binary input. Is there a red item in the laundry? No. Are there black items? Yes. Are they nylon? No. Are they cotton? No. Are they silk? Yes. Do these black silk items have an existing stain? Yes. Is it a stain from coffee? No. Is it a stain from mayonnaise? No. Is it a stain from cheese sauce? Yes. And so on. These binary decision trees can grow to millions or even billions of nodes for complicated subjects. The algorithm's tree may take inputs and run them through equations and formulas, then take those answers as further inputs, creating long strings with repetitive layers and mind-boggling detail. The German mathematician Gottfried Leibniz theorized on this exact subject three hundred years ago—that life could be broken down into a long series of binary decisions—long before there existed semiconductors to facilitate machines that could run algorithms.

Peterffy's algorithms grew to be spidery matrices with hundreds of inputs, variables, and dependent differential equations and integrals. Putting such things into computer code took a true master and a large team. At Mocatta's headquarters in Manhattan, Peterffy's programmers worked at computer screens and read market data as it came in on Teletype machines. The programmers then typed the data by hand into their computers, whose algorithms issued prices for Mocatta to quote on the New York Commodities Exchange floor. The programmers, speaking to clerks near the floor action downtown, would bark out quotes as fast as the algorithm issued them, and Mocatta's clerks would signal the prices to their pit traders with hand gestures. It was hardly high-speed trading, but it was the first time markets were consistently dictated by an algorithm. And the best part for Peterffy: the rest of the market had little idea where he got his numbers from.

He might have entered Wall Street as a hacker with little market sense, but Peterffy's trading instincts—born while chopping up chewing gum sticks, chasing scrap metal, and pawning stamps—became sharpened. Jarecki noticed. He came to lean on Peterffy for advice on any significant decisions or trades that Mocatta made. One coworker remembers Jarecki never entering an important meeting without Peterffy at his side.[2]

By 1976, Peterffy's hacker force had grown to eighty. It was the largest financial programming operation in the world. Peterffy had proven himself not only a nimble writer of code but also a capable manager of disparate and sometimes brilliant scientific personalities—a rare and nuanced skill.

Even as Mocatta grew, its business remained mostly in commodities. Peterffy wanted to bring his algorithms to the stock options pits, but Jarecki wanted to remain focused on metals. The markets for stock options were dominated by traders who had come from the stock exchanges, where a company's value was determined by men in the pits—and rightly so. There are a number of nonquantifiable elements affecting stock prices, such as goodwill, growth prospects, looming lawsuits, and competition; for that reason, there's no magic formula for determining the price of a stock. Option prices, however, should be pure reflections of probability and statistics. Only a very select few, including Peterffy, knew this. For those who understood the math, opportunity ran like a river.

When the Chicago Board Options Exchange became well established by the late 1970s, Peterffy took it as a sign that the stock options market was about to explode. In 1976, he traveled to Chicago to check out the CBOE. The bid-ask spreads on some of the options were as much as two and three dollars. "The traders just made these prices up and threw them out there," he says.

With the burgeoning possibilities of stock options, Peterffy wasn't satisfied sticking to commodities. Volume on the gold and silver markets and their related options, Peterffy knew, weren't sufficient to build a true Wall Street fortune. He was also thwarted in his desire for a piece of ownership at Mocatta, which he says Jarecki had promised him. Jarecki denies ever making any such agreement.

In 1977, Peterffy spent $2,000 on his first home computer, an Olivetti. After putting in days at Mocatta, he spent nights programming his Italian machine with algorithms, preparing for the day he would storm the market for stock options.

THE ORIGINAL ALGORITHMIC TRADER

With no possibility of becoming a part owner at Mocatta, Peterffy decided he had squeezed all he could from his tenure there. So in 1977, with $200,000 in savings, he left the stability of employment and bought a seat on the American Stock Exchange, which had just begun to trade options. The seat cost him $36,000, leaving him $164,000 to trade with.

Leading up to his first day at the AMEX, Peterffy toiled at his computer for eighteen hours a day, tweaking his algorithms and producing sheets to guide his buying and selling of options on the floor. He homed in on a few dozen companies whose options he believed were more often mispriced than others. For each company, he made a series of sheets listing option values for different prices of the company's stock. As stock prices fluctuated throughout the day, Peterffy planned to quickly consult his sheets, determine the fair price of the stock's options, and buy or sell accordingly. The sheets were the paper manifestation of his algorithms.

Peterffy packaged all of his papers together into a three-ring binder that would serve as his portable cheat sheet in the trading pits. He found preparing for the AMEX difficult, not because he wasn't confident in his calculus—his methodology was his rock—but because he was scared of working alone, being utterly independent. He was worried about his accent and having to jostle, bounce, and shout with throngs of demonstrative and sweaty men in the pits. He fretted about clubby groups of traders ganging up on him, ending his experiment prematurely and unfairly. Most of all, he worried about failing, about having to return, defeated, to Mocatta.

Determined to not let that happen, Peterffy showed up on the AMEX floor, binder in hand. He quickly discovered that the binder, when held out in front of him, effectively made him the size of two traders in the packed pits. Other traders did not appreciate surrendering space so that he could crack open his magic book. And just what the hell was in the book, anyway? the other traders asked. Trading, they preached, was about your wits, your guts, and your balls, not some goofy clutch of papers.

"People thought I was absolutely ridiculous working out of this binder," Peterffy says.

So to fit in, and to make life generally easier for himself in the pits, Peterffy boiled his most important notes down to a series of double-sided sheets that he could easily fold and stuff into his back pocket. When he thought the price was right, he would dip into his pocket and consult his sheet. As he stood there reading the tiny print that came from his algorithms, the tide of the pits shoved him to and fro. But he carried on, oblivious to the jostling, until he could shoot his head up, raise his hand, and catch the eye of another trader or market maker, barking out his order in a heavily accented burst of words. The accent, in fact, was a problem; so too was his peculiar habit of consulting grungy folded sheets of paper from his pocket. The other traders, Peterffy says, "thought I was quite mad."

Peterffy would sometimes spend hours in the pits without uttering a word or making a single trade. He religiously consulted his sheets. If an option didn't fit his very conservative profit guidelines, he didn't buy it. "I was very careful," he says.

Despite his caution, Peterffy couldn't avoid the market's mandatory beatings for neophytes. Early in his trading career, he spent a morning swapping options on DuPont. In the middle of the session, he noticed an out-of-the-money call option selling for $31. According to his sheets, the option was actually worth $22. He planned to go short. "That was a big profit for a little Hungarian guy," he says.

Peterffy actually had three hundred of the options contracts that he had picked up for $18, so he sold those at $31 and then sold two hundred more. He had sold two hundred call options without covering his call—he hadn't bought a corresponding number of shares or hedged his bet with put options in the other direction. A move against his bet could crush him. But how could he pass up such a deal?

Just after Peterffy sold the calls, DuPont stopped trading. News trickled in. The company announced earnings much larger than expected and a two-for-one split of the stock, whose price soared. Peterffy's two hundred options contracts, each representative of one hundred shares of stock, left him on the hook for a $5 loss per share, or $500 per contract. He had just blown $100,000, more than half of his trading capital. He blames the incident on people trading with insider knowledge. Whatever the reason, the $100,000 loss was a calamity. "I didn't know if I should throw up or cry," he says. He went home that night devastated, unsure of his methods. "I had thought I was so clever."

Smoking a cigarette, staring at his small kitchen table in his Upper East Side rental, Peterffy thought he needed to dedicate even more of his life, his money, and his attention to trading, although by most standards he was already maniacally focused. The only way to dig out of this hole, he decided, was to become a machine bent on one thing: trading, saving, and winning. Nothing else mattered. Peterffy got up from the table and threw his pack of cigarettes into the kitchen garbage. He would never smoke again. The cash he spent on cigarettes would be better utilized on the trading floor, he reasoned. "I instantly calculated how much money I could save in twenty years by no longer smoking," he explains. "I needed everything."

Peterffy returned to the pits with a renewed focus. He stuck to his sheets, as always, but with DuPont haunting him, he didn't make what he called "cowboy bets."

He slowly rebuilt his capital, one grinding day at a time. Sticking to his algorithmic system, he rarely experienced days with substantial losses. Even though the Black-Scholes formula had been published seven years before, it wasn't moving the markets enough to bother Peterffy or others who were cashing in on its genius.

As effective as his algorithms and sheets were, Peterffy was only one man. He needed more people in the pits. So he slowly hired more traders. To prevent losses and keep control of how his traders operated, he trained them to bid and offer only off of values on his sheets, which he would update with fresh numbers from his algorithm every night. As he expanded his operation on the AMEX, Peterffy renamed his trading operation Timber Hill, after a road in rural New York where he vacationed. He bought more trading seats and began dabbling in other strategies such as pairs trading and arbitrage, both of which he was familiar with from his time developing tactics for Mocatta.

Even with additional traders jostling in the pits for him, Peterffy spent hours ruminating on ways he could drop all of his floor-trading theories into a computer that could execute his algorithms far better than people on a trading floor. But there was no avenue to do such a thing, not yet. The nature of the pits bedeviled him. The ceilings of the trading floor were more than forty feet high and the pits themselves were lined by railings where clerks would stand, far above, and signal to their traders below what to buy or sell. "Everybody knew everybody's signals," Peterffy says, so that much of the pits knew what was coming from some traders before they ever lifted their arms, resulting in their trades being ambushed or stolen. "It was ridiculous."

Peterffy circulated an idea that bids and offers could be entered by all traders on handheld devices tethered to a central computer that would automatically track price-time priorities and spread the deals out fairly. The idea didn't last long, as those in control—the specialists—quashed it.

Rather than changing how the market worked for the better with technology, Peterffy decided to join in a larger capacity and become a market maker in some options on the AMEX, thinking the higher volumes would lower his risk and raise his profits so long as he was able to keep trading in bands he knew to be fair according to his algorithms. But first he had to find a way to have his bids and offers instantly and consistently recognized by the specialists who controlled trading. Because he didn't traffic in the normal scuttlebutt and sports chatter that most traders did, he was something of an outsider. The specialists didn't always call out his trades from the crowd because they simply didn't enjoy dealing with him. So the Hungarian decided to hire people whom he knew the specialists would like.

The financial industry, as is the case with most high-paying fields, tends to be dominated by men who are wont to hire more men. So when Peterffy hired the tallest, prettiest, most buxom women he could find, the plan was more than a bit novel. The tactic worked miracles for his order flow. Suddenly, the specialists always took his trades. They put their arms around his traders, chitchatted, and recognized the blondes' orders as fast as they were issued. "The specialists were thinking, 'These dumb blondes, what do they know, right?'" Peterffy says.

It's true that the women Peterffy hired didn't know much about trading, let alone algorithms. But none of his traders at that time were any good out on their own. And none of them were using the sheets for guidance anymore. Peterffy had devised a new system that empowered anybody to make smart trades.

Like many innovations, the system was conceived by accident. In the middle of 1982, Peterffy tore several ligaments in his knee. During his rehab, the knee became infected and he was unable to stand on the floor for long periods, let alone wrestle with the meatheads in the pits. Relegated to his office upstairs at the AMEX, he called trades downstairs and was left by himself to tinker. His attention eventually settled on his Quotron machine, which gave users one stock or option price at a time. The data for the Quotron came in on a dedicated phone line. Peterffy had asked Quotron, which dominated the data industry at the time, if they would sell him the feed they sent to their machines. The answer, every time, was a curt no.

Cooped up in his office, Peterffy took to the Quotron feed with a pair of wire snippers, a move that would later inspire his Nasdaq hack. "So, of course," he says, "we stole the data."

Upstairs, having severed the wire to his Quotron machine, Peterffy wielded the old-school tools of an electrical engineer. He used an oscilloscope to measure the electric pulses in the wire and, matching the pulses to data, decoded what the wire's signals meant. Once he knew that, he wrote a program for his PC that would inhale the Quotron data as it was issued and store it within its memory. The program would then scan the warrens of stock and option data and route it through Peterffy's algorithms. First and foremost, the algorithms searched for options that were egregiously mispriced.

What Peterffy was especially interested in—and what he wrote his algorithms to search for—were what's known as delta neutral trades. In these trades, an overpriced call option, which Peterffy would sell, could be coupled with buying an underpriced put option to create a position that wouldn't be adversely affected by a spike or a dip in the market.

For example, say shares of IBM were trading hands at $75. A delta neutral trade for Peterffy could look something like this: He would sell 10,000 call options (the right to buy IBM at a strike price of $75 during the next sixty days) for $1 each (the overpriced option), taking in the $10,000. Almost simultaneously, another Peterffy trader would buy 10,000 put options (the right to sell IBM at a strike price of $75 during the next sixty days) for 85 cents each (the underpriced option), for $8,500. The result of the trade: a near-riskless $1,500 profit.

Because options on stocks of big companies such as IBM traded at dozens of different strike prices as well as multitudes of diverse expirations, the 1980s options market was rife with delta neutral trades for those who could find them. Some floor traders had caught on to the tactic—they scanned the tapes and tickers for mispriced options that could easily be hedged for riskless profits. But the power of a few men searching out trades in the pits was no match for Peterffy's indefatigable machine.

The computer would put together trades that made for delta neutral plays and print them out immediately. Peterffy, confined to a chair with a bum knee, would then place a call to his floor clerk, who would get the trades to his corps of women. And on it went. Four months after hacking the Quotron link and getting his computer system operational, Peterffy was making more money than ever. The key to it all was a dependable flow of pure data that few others had. And data, as so many hot companies of today have demonstrated, can be the difference between domination of an industry and failure. Peterffy's operation pioneered the automated compilation and employment of vast data stores on Wall Street, where the mining of such things got its start.

THE ALGORITHM GOES HOLLYWOOD

After installing a trading system that leveraged a hacked Quotron machine along with attractive female traders fresh to the pits, Peterffy felt he had built the perfect profit engine. "Anybody," he told friends, "can make money trading with me. Anybody."

Wall Street men being Wall Street men, many called Peterffy on his assertions. How ridiculous his bragging must have sounded—to claim that anybody in his system could wade into a pit full of veritable experts, people who had spent entire careers honing their instincts and trading prowess, and take money from them. Peterffy stood his ground, and to prove it, he brought on Melvin Van Peebles, who was friends with Peterffy's old boss, Jarecki, and as unlikely a person as any to end up in the trading pits. In 1971, Van Peebles wrote, directed, produced, and starred in *Sweet Sweetback's Baadasssss Song* in a furious nineteen days. He needed a $50,000 loan from Bill Cosby to get it done, but the film eventually became a hit, grossing $10 million and launching Van Peebles's career as a producer, director, and actor.

Despite his celebrity status and an already busy life (at the time, Van Peebles was busy producing Broadway musicals), the writer, intrigued by the nature of Wall Street, went to work for Peterffy's Timber Hill alongside the firm's other traders, all women, in November 1982. He remained on the floor for Timber Hill an entire year, surprising almost everybody except Peterffy and himself. "I'm doing just what I've always done—making deals," Van Peebles told *New York* magazine.[3]

Van Peebles also made money, a lot of it. He explained his success: "You gotta be able to calculate doo-boop-be-deeliyaboop—deal! I can do that."[4] He had learned math, he said, studying astronomy in the Netherlands and in the air force, but it's most likely any math being done for his trades came straight out of Peterffy's algorithm, which was gnawing on raw data from the hijacked Quotron line.

Just as with Peterffy's other traders, Van Peebles frequently ran sorties to a large bank of phones on the trading floor to communicate with Timber Hill's headquarters. On the phone, he scribbled down a jumble of letters, numbers, and fractions—his instructions. From there, he careened back into one of the roiling pits and put up his hands, ordering fresh delta neutral trades. Van Peebles's story accentuated the success of the most improbable trading squad roaming the pits of New York, perhaps to this day: three blonde women and one highly acclaimed black writer, director, and actor, all of them well-disguised proxies of an algorithm that dwelled inside a machine.

THE IPAD'S FORERUNNER

Peterffy's strategy to play to the pits' infamous chauvinism paid off as the options specialists continued to take his women's orders above most others. But after six months of his generating easy profits, the specialists began to notice the uncanny market sense that Peterffy's traders seemed to possess.

One of the specialists came to Peterffy and said, "Look, we know these trades are no good. We're getting killed on every single one. What are you doing?"

Keen observers assumed Peterffy had figured something out, but they couldn't be sure what it was. What they were certain of, though, was that they wanted little to do with the trades Peterffy's women were bringing to them—as much as they might have liked the attention. Sometime before this, Peterffy had been designated a market maker by the specialists, which gave him first crack at new orders in the pits. Technically, a market maker is required to keep both bids and offers up at all times, no matter where the market goes. But Peterffy had been bending the rules, as a lot of market makers did, cherry-picking the trades he wanted according to the instructions of his algorithm. At no point was he maintaining constant bids and offers.

The specialists, sick of getting beaten by Peterffy on nearly every trade they did with him, told the Hungarian that he would have to maintain open bids and offers on a minimum number of options or else they would pull his market-maker status. He quibbled with the demands as much as prudence would allow, while his brain searched for a solution that would keep profits flowing and pacify those who might scuttle him.

Because he was required to keep constant bids and offers up on certain securities, Peterffy couldn't ask his traders to duck into the phone bank for every directive. Keeping quotes open at all times meant sticking to the pits and paying attention to market movements with vigilance. How could he expect his traders, who largely took their directions straight from an algorithm housed in the guts of a personal computer, to truly make markets without courting disaster?

The answer was buried in one of his past suggestions to the exchange: handheld computers. Peterffy had previously pushed to outfit the whole floor with the devices to do away with the anachronistic methods of the specialists, but he had made this proposal without a specific device in mind (they didn't exist) and without properly considering the dogmatic ways of the exchanges and who truly held power: the exact people he had suggested be displaced.

Peterffy's new plan wasn't to equip the entire exchange with handheld computers—just his traders. But would such things even be allowed? Peterffy hoped so; if he could succeed with his devices on the AMEX, he planned to take his game—and his algorithms, computers, and pit machines—to the Chicago Board Options Exchange, the largest options trading floor in the United States.

The AMEX governors weren't crazy for Peterffy's pitch; some members were hysterical about the matter. Machines, on the floor, trading!

Most objectors claimed the machines would get in the way of the jostling traders and perhaps even issue reckless orders that Peterffy wouldn't be able to cover if the market soured on him.

In spite of the protests, the AMEX agreed to allow Peterffy's traders to bring small tablet computers into the pits. Peterffy now had a new problem: he had no tablet computer of any kind. Nor did anybody else, thirty years before the advent of the iPad. The devices would have to be built from scratch.

With a vague idea of what he wanted, Peterffy brought in physics PhD candidates from NYU who helped him construct small rectangular boxes out of Mylar, a black plastic. The boxes measured roughly eight inches by twelve and were two inches deep. Inside were packed transistors and circuit boards that gave way to a top panel that included a series of gold wires. The wires detected when users pressed them with a finger, creating a touchscreen. Peterffy could then drop different sets of thin plastic templates on the boxes to serve as keypads. Different kinds of options required different keypads and different programs. Each overlay was similar to an app. This way, Peterffy could make just one type of box and then program each with the algorithm for what its user would be trading.[5]

In the morning before the market bell rang, Peterffy would crack open each device and pluck a small wire harness from inside, plugging it into the PC that had been collecting data from the Quotron line. From the computer, the handheld device would download the latest market prices and data, enabling it to instruct its pit-bound user on where to quote stock and options prices. Users punched in current market prices, and lights on the tablet's display indicated whether the trade was worth making. Now when specialists called on Peterffy's women for a bid and offer, the women could quickly give prices, knowing they were aligned with Petertffy's algorithm.

At this point, Peterffy was making more than $1 million a year. The question that faced him now was, how big could this become? Having tamed the AMEX with algorithms, he looked for new challenges. He felt confident he could dominate any market to which he could apply his code. He set his sights on the Chicago Board Options Exchange, where volumes were biggest. The governors' board there fought him and refused to allow his handheld devices on the floor, claiming the boxes were too big and wouldn't fit inside the crowded trading pits. Peterffy accepted this and went back to his workbench, fashioning new, smaller boxes that measured seven inches square, a big reduction on the original design. Presented with the new boxes that could be easily held tight to a trader's chest, the CBOE cut straight to the heart of the matter: members of the exchange would tolerate absolutely no computers of any kind of the floor, clearly afraid of what kind of edge the devices might afford their users. Benefits to investors and the market in the form of efficiency and perhaps tighter spreads wasn't a consideration.

"Of course the exchanges fought me," Peterffy said. "They always fought everything."

So Peterffy turned to the New York Stock Exchange, which wanted to bolster its position as the dominant equity exchange with an options floor. Normally a kingpin that dictated terms, the NYSE was eager to court Peterffy's volume. To further assist his traders, who were already armed with pit computers, Peterffy constructed a set of light bars he mounted high above his clerk's post on the floor. The bars were broken up into a series of different-colored bulbs. He then wrote a system in which he could signal new trades to the pit using the colored bars.

Everything still relied on his computers' programming, which had recently been overhauled and rewritten to operate in C, a more modern and efficient programming language than the Fortran that Peterffy had used for most of his shop's code. As always, the computers combed incoming data for ripe trades. When the computer struck opportunity, it would immediately send a series of electric pulses to the colored bars on the NYSE floor, which a set of Timber Hill's traders watched from opening bell to closing. Before long, Peterffy became one of the major market makers at the NYSE.

Since buying his original seat on the AMEX for $36,000 in 1977, Peterffy had kept his main offices at that exchange. With further expansion on his mind, he needed more room than the AMEX could give him. So in 1986 he moved his headquarters to the World Trade Center, where he had more room to command traders at multiple exchanges.

Moving operations to the Trade Center, which was several blocks from the trading floors, made it harder to keep the tablets updated with fresh data, because the PCs that supplied the data had been moved as well. Peterffy's simple solution to this problem gave an unwitting nod to Wall Street's past, when runners used to broadcast word of what was happening on the markets to people uptown before the news could spread otherwise. He hired a couple of swift workers whose main purpose, outside of sundry office work, was to sprint from the World Trade Center to the exchanges with updated handheld devices under their arms.

"If you were ever downtown and you saw some crazy guy running as fast as he could down the street with a black box," said Peterffy, "then you saw our guys."

The running men eventually lost their jobs to dedicated phone lines carrying data, which were beginning to find homes in some of the tech-savvy trading houses by the mid-1980s. Peterffy was at the head of the line, leasing phone lines that fired data back and forth between his Trade Center offices and his exchange outposts, where computers would consume data from headquarters and dump it into the handhelds. Even with this update, traders still had to run the devices from the pit up to the computers at the exchanges several times a day. To eliminate this step, Peterffy and his engineering team built mini radio transmitters into the handhelds as well as Timber Hill's computers at the exchange. With that task completed, data now flowed effortlessly to the handhelds on the trading floor.

By 1986, the pits had become cash machines for Peterffy's trader corps. The algorithms' orders flowed out, money flowed in. The radio-enabled devices gave Peterffy the freedom to move more volume without the fear of posting a quote that was out of line with his algorithms. Timber Hill started 1986 with $1 million in capital. It ended the year with $5 million in the bank, a 400 percent return. The successes, now piling up, allowed Peterffy to think expansion.

As technology advanced, so did Peterffy's most adroit opponents, who copied his methods. Joe Ritchie's Chicago Research and Trading Group had become a power in Chicago, and Blair Hull's Hull Trading charged up the ladder quickly after being founded in 1985. O'Connor & Associates, also in Chicago, was employing very similar tactics to Peterffy's, outfitting its traders with cheat sheets for valuing options and supplementing that information with computers that constantly crunched data upstairs while piping new numbers down to the pits. O'Connor was so secretive about its methods that when it bought two hundred Symbolics computers in the mid-1980s, executives shredded the packaging so Dumpster-diving competitors couldn't determine what technology the firm used.[6]

THE ALGORITHMS SPREAD COAST TO COAST

By 1987, index funds, which tracked groups of stocks such as the S&P 500, had grown popular not only with the public but also with professional traders. But certain indexes, the S&P 500 included, could only be licensed for trade in one market. In the case of the S&P 500, the license belonged to the Chicago Mercantile Exchange. So other exchanges employed indexes that were not exactly the same, but close. The Chicago Board Options Exchange traded the OEX, which was the same as the S&P 100; the New York Stock Exchange traded the NYSE composite, which tracked the entire NYSE; the AMEX traded the Major Market Index, which tracked the thirty biggest stocks; and the Pacific Exchange had what it called the PSE, which was based on technology companies, a growing share of the market.

All of these indexes contained different things, but their core holdings were similar. The S&P 500 index, although only five hundred companies, comprised 90 percent of the weight of the NYSE index, which contained every company traded on the Big Board. Because the five hundred largest stocks on the Big Board were so disproportionately larger than the rest of the market, the NYSE index basically tracked the S&P 500 index and vice versa. The same could be said for the S&P 100, the Major Market Index, and the rest.

If the indexes more or less followed the same path, Peterffy reasoned, then their volatilities were the same and so should be the price of their options and futures. In reality, however, the prices of these instruments could vary greatly from exchange to exchange. A November call option on the NYSE index might be trading for $2 in New York, while a similar call might be trading for $3 on the OEX index in Chicago, and a third one for $2.25 on the Pacific Exchange in San Francisco. "So it was fairly obvious, at least to us, what people should be doing," Peterffy said. This was child's play: sell the expensive index derivatives and buy the cheap ones. "It was a wonderful thing," he recalls fondly.

To take advantage of all these easy trades, Peterffy needed people on the floor in San Francisco, the two Chicago exchanges, and the two New York exchanges. He and his team of tinkerers manufactured new sets of handheld devices for the waves of trading recruits who would establish Timber Hill beachheads on the country's other trading floors. Peterffy bought dozens of new computers and leased cross-country phone lines that would stay open for data at all times, allowing his computer network to maintain real-time contact and adjust his team's prices on the fly across the entire country. Radio transmitters were fitted and installed at the new exchanges, giving Peterffy's Chicago pit traders the same information available to those in San Francisco and New York. Now a sale in New York could be instantly hedged with buys in Chicago.

When the values of the derivatives converged back to their expected ranges, Peterffy's computers would tell his traders to dump both the long and the short positions, locking in profits. Because the index derivatives were heavily traded in all markets, there were hundreds—if not thousands—of arbitrage opportunities every day. Few, if any, traders were taking advantage of technology the way Peterffy did. Some trading houses kept open phone lines between New York and Chicago so that clerks could bark prices back and forth and pounce on large pricing discrepancies. Peterffy's automated system allowed his traders to harvest not only large mispricings but also smaller ones—and they almost always got to them before others. Peterffy had created the first algorithmic trading operation working from coast to coast.

All trading activity from the handhelds was radioed to waiting terminals Peterffy had installed at each exchange. The computers there would then wire the data across the leased phone lines straight to Timber Hill's offices in the World Trade Center, where it would be received by a large master algorithm called simply the Correlator, which ran phalanxes of code to dissect markets and pinpoint their weaknesses, while dispatching Timber Hill traders in each city to hammer them. The Correlator analyzed real-time prices across a dozen equity and derivative markets and issued salvos of trades almost guaranteed to generate easy profit. As completed trades came in through the data lines, the Correlator swallowed Timber Hill's positions and spit out trades to hedge them. Some of the Correlator's trades were 100 percent automated—they were shipped directly over to Peterffy's hacked Nasdaq terminal, which rapped the keyboard automatically.

A trade that began with the Correlator and ended with an automated execution through Peterffy's co-opted Nasdaq terminal constitutes phase two of algorithmic takeover. Here algorithms comb the data, read the market, and issue an order that is no longer carried out by humans but by another machine. What Peterffy ultimately helped create was a Wall Street whose most important communications travel back and forth between machines, not humans. It would be more than a decade before this second-degree takeover spread across the entire U.S. market system, but it all started with Peterffy and the Nasdaq. After Peterffy's trick here, there remained only phase three—wherein algorithms adjust independently of their human masters, and in some cases write algorithms of their own—to complete a total bot takeover.

Peterffy often sat in his World Trade Center office watching his traders' work pour into the Correlator. The screen listed completed trades and how those trades should be hedged, sending word to Timber Hill's people in the relevant markets. Once the hedging trades had been made, they too would be posted on the Correlator's screen. Peterffy had never been a gambler. His game was to make easy bets where profit, albeit sometimes small, was assured, while keeping the downside as close to zero as possible. When large trades popped up on the Correlator's screen, Peterffy would focus intently on the monochrome pixels, watching for his traders to hedge the bet. But even the mighty Correlator had its weaknesses—as do all algorithms left on their own with no human supervision.

One morning in early 1987, as Peterffy watched trades stream in, he noticed a whale. One of his traders sold 100,000 NYSE index puts, giving somebody in the crowd the right to sell 100,000 shares of the index at a guaranteed price in the future. If the index were to crash, as it would eventually do later in the year, such a position could be ruinous for Timber Hill. Seeing the trade, Peterffy's brow arched as he wondered who would be making such a large and bearish speculation. His mind, as it often did, wandered back to his disastrous DuPont trade. He had failed to hedge his DuPont play with a bet on the opposite side, which was why it had proved so calamitous. But now Timber Hill hedged everything.

Peterffy sat up in his chair and waited for what he knew was coming: the hedging trades. Just like it was supposed to do, the computer surveyed the markets and determined the cheapest way to shed the trade's risk and dispatched orders to do so. Because selling the puts was a bullish trade, the Correlator ordered traders to buy similar, cheaper puts on other exchanges as well as selling indexes similar to the NYSE short. His system worked exactly as he had programmed it. Peterffy eased back into his chair as the hedged trades were confirmed.

But then two minutes later the large trade popped up again: a trader sold 100,000 NYSE index puts. Ghosts of DuPont were poking Peterffy now. He whirled and picked up the phone, dialing his traders' desk at the NYSE.

"Who are you selling these puts to?" Peterffy said.

"What puts?" came the response. Timber Hill had six traders on the NYSE floor, so it was perfectly understandable that the trader who answered the phone didn't know about the sale. Peterffy explained the trade. "Go find out who is doing this," he said, hanging up.

Now ruffled, Peterffy sat back down. He watched the Correlator mete out more hedging moves, which his traders duly carried out. Then his stomach turned. For the third time, the Correlator screen read, SOLD: 100,000 NYSE INDEX PUTS.

"What is this shit?!" Peterffy yelled. He scrambled back to the phone and dialed the NYSE floor. A trader picked up.

"What the hell is going on—who is selling these puts?" Peterffy said.

"I don't know anything!" the trader responded.

Peterffy yelled, "Pull out all of the cords, just pull them out!"

The trader immediately turned around and pulled out all of the power and data cords going into the exchange computers, shutting down Timber Hill's operations on the NYSE. Their handhelds not receiving any information, Timber Hill's remaining NYSE traders, confused, wandered back to the office.

Peterffy rushed out of the World Trade Center and into the New York Stock Exchange. Scaling the stairs at the neoclassical NYSE headquarters at 11 Wall Street, he swung open the door to Timber Hill's small office. He interrogated all of his traders. "How is it that nobody has any idea where these trades came from?" he asked.

His mind jumping, Peterffy decided to count all of his handheld devices in the building. Each trader quickly produced their tablet. They were all safe and accounted for. But then Peterffy remembered there was a spare device kept in the office in case another broke, something that happened fairly often. Peterffy quickly spotted the small rectangle, which sat on a desk near the door. Just as he went to retrieve it, one of his traders, who had been in the bathroom, came back into the office. A noticeable whoosh of air accompanied him as he opened the door to the office. Peterffy looked at the device, whose plastic overlay riffled as the air passed over it.

The door to the office closed and the air grew still. But then the door, forced by the air outside the office, cracked open again, sending another burst of air across the tablet, making its touchscreen overlay crinkle out loud.

"Turn on the computer," Peterffy, now intrigued, said to one of his traders.

The door cracked again, and the computer, now on and registering the tablet's entries, showed a sale for 100,000 NYSE index puts. The small blast of air from the door applied enough force to the device's face that it registered sales as if somebody were entering them with their fingertips. The order kept coming through because the office door was being continuously bumped open by air rushing in from the trading floor to the hallway. For a trade to be registered as legitimate, a trader had to not only enter it but also follow up with a confirmation. When the tablet blinked for the confirmation, the air feathered the yes button on the tablet and the Correlator was notified of the trade.

None of these big trades had ever happened. But the Correlator, working as it was designed, didn't know that. All of the hedging trades it ordered up weren't hedging at all. They were straight trades, pure and simple. When the market closed that day, Timber Hill was sitting on more than $3 million worth of naked directional bets. Peterffy had just become what he worked so hard to avoid—a high-stakes betting cowboy. He would have to wait until the next morning to unwind his positions; he simply had to hope the market didn't move against him before the opening.

That night, Peterffy couldn't sleep. He stared at the ceiling all night long until it was finally time to go back to the office.

Luckily, the market hadn't moved much and Timber Hill was able to dump its positions with little bloodshed. Had this been a later day in 1987—October 19's Black Monday—these erroneous big bets could have ended Peterffy's career.

A HACKER RISES TO ROYALTY ON WALL STREET

In 1988, Peterffy made $50 million, an astonishing amount of money to him and a sum that marked his arrival to the big time. From his lead vantage point—few people knew what he was up to—he sprinted out ahead of the pack and continued to dominate stock and options markets with bots that consistently outdueled their human adversaries. At some point, however, the rest of Wall Street began to catch up—and that's when quant mania took hold of lower Manhattan and began to beckon scientists and engineers to careers in finance.

In 1999, Goldman Sachs offered Peterffy $900 million for his business. He wanted $3 billion. Goldman instead bought Blair Hull's automated trading operation in Chicago for $500 million. Hull's shop was far smaller than Peterffy's, but Hull had made waves when, more than six years after Peterffy started the trend, he began using bots to trade on the Cincinnati Stock Exchange in the early 1990s.

It was only a matter of time before the investment banks and Wall Street's blue-chip players started to catch up with Peterffy and the other programmers who had been getting the better of human traders for years. The monsters of finance hired their own hackers and engineers and moved into the space as quickly as they could. Hull ducked out of the game when he sold to Goldman, later losing a U.S. Senate primary race to Barack Obama in Illinois. Other algorithmic pioneers left even earlier. O'Connor & Associates, the shredder of computer boxes, sold to Swiss Bank in 1992, and Joe Ritchie's Chicago Research and Trading Group sold out to NationsBank, which would later become Bank of America, a year later. Peterffy, the original algorithmic trader, remained to face off against the self-appointed masters of the universe. The Hungarian's operation, which he eventually renamed Interactive Brokers while moving the company's headquarters to Greenwich, Connecticut, continued to pace Wall Street in the 1990s, as it does to this day.

Part of the reason for this was that Peterffy's organization was, and will remain, he asserts, one of engineers and programmers. Goldman and the others on Wall Street seek out engineering talent to stock their quant departments, but at Interactive Brokers, the engineers *are* the company—making Peterffy's firm something akin to Wall Street's version of Google, a place where engineers make the product and the big decisions. Interactive Brokers, in fact, works to ensure that 75 percent of its employees are programmers and engineers. "Most Wall Street firms concentrate on what they do best," Peterffy says. "And that means they sell. But we write code. That's what we do."

Nobody at Interactive Brokers has a business degree, Peterffy points out, making the firm flagrantly out of place among its peers. Peterffy insists he won't have an MBA at his company—ever.

During the 1990s and into the 2000s, Interactive Brokers expanded, bringing its algorithms to European markets as he continued gaining more share in the United States. On Friday, May 4, 2007, Peterffy prepared for a day that all companies, tech and finance alike, dream of having. It was IPO day for Interactive Brokers. He wore a beige suit with a subtle check pattern paired with a pressed white shirt and a tie so blue it nearly looked purple. He arrived at work early, sipped green tea, and peered out the corner windows of his office at the bubbling traffic of Greenwich.

By 7:30 a.m., Interactive Brokers was humming. Peterffy headed out of his office and greeted visitors. At 9:29 a.m., several dozen people all grouped into one place, Peterffy at their center. As the clock hit 9:30, he raised his hands, clapped, yelled, and brought his hand quickly down on a large orange button as applause erupted all around him. Twenty years after the Nasdaq had tried to dismantle his automated trading operation that subsisted on spliced wires, hacked data lines, and custom-written code, Peterffy rang the Nasdaq's opening bell. The market valued his company at $12 billion.

Peterffy retained an 85 percent ownership share of Interactive Brokers and only put 10 percent of its stock up for sale in the offering. Even so, the deal put $1.18 billion directly into his pocket and was the second-largest IPO in the United States that year. Not a bad outcome for a Hungarian immigrant who, lacking a full engineering degree, decided to pick up programming by reading computer manuals in the 1960s.

THE UNKNOWN FINANCIAL FRONTIER

The speed and volume of trading continue to tick up on Wall Street as fresh code and newer and faster hardware flood the game. As the arms race played out on Wall Street in the late 1990s and into the 2000s, some of its methods and talents began to leak into other fields. Wall Street may have cribbed some of the science, but no other industry has so perfected the development of endlessly complex algorithms and automated bots.

In today's stock market, humans have largely been reduced to interested observers. The algorithms own the market now. Dow Jones and Bloomberg offer news services that are written specifically for the trading bots. These stories would be incomprehensible to a human, but make perfect sense to a bot. The news affects the markets, and just as Wall Street traders once skimmed *Barron's* on their train ride in from Greenwich, now algorithms read the paper too. They just read it a lot faster.

When machines are at the controls, we can never be sure exactly what the market will do. The disappearance of $1 trillion in wealth on May 6, 2010, is a case in point. And the aftershocks continue. Six months later, Progress Energy, a utility in North Carolina, saw its shares shed 90 percent of their value in minutes for no apparent reason. Apple shares, which rode wild price swings on May 6, inexplicably dropped 4 percent several months later, wiping out $16 billion of market cap before rebounding. Michael Kearns, a professor at the University of Pennsylvania who has written algorithms for traders and studied this new, machine-ruled market, says there's no way to understand the implications of giving algorithms the full control that they now have.[7]

The competition between Wall Street algorithms has become so bizarre that there are days when 40 percent of the trades on all U.S. exchanges, from the Nasdaq to the NYSE, are made by just two midwestern companies that most people, even those who work in finance, have never heard of. One of them, Getco, is located in Chicago; the other, Tradebot, is in Kansas City. Both firms employ world-class hackers and engineers who are focused on clearing profits of often less than one cent per share. Getco and Tradebot deploy thousands of algorithms to scour the markets for the tiniest of opportunities. And there are thousands of other companies, some big, some small, that exist only to do the same thing—to make money on the market with bots. All in all, this means that algorithms *are* the stock market, as they now carry out 60 percent of all trades in the United States; European and Asian markets aren't far behind. What once was determined by biting, swarming, and barking hordes of men is now decided by battling algorithms that are constantly testing each other, grappling for an edge, feigning their intentions, and learning as they trade.

There's plenty of good that has arisen from algorithms' role in our markets—the normal man's ability to trade stock for $7 from the laptop screen in his kitchen being chief among them. As the market shifted from one where each order had to pass through pits of humans to one that's facilitated by algorithms and exchanges hosted on computer servers, the costs of trading have plummeted. But we've blown by the utility of cheaper traders to a strange Star Wars–like battle that features fewer human inputs and more algorithmic autonomy. The speed at which something can go wrong is frightening.

Our stock markets have become such specialized battlegrounds that some algorithms sit silently for months, waiting to ambush enemy algorithms that display design flaws, dated code, or a discernible pattern to their trading. Many of the algorithms plying our stock markets are built to mimic randomness. What's random can't be gamed, planned upon, or hijacked. For example, some of the most important algorithms on Wall Street trade stocks that belong to normal people—people who may own mutual fund shares in their 401(k) plans or IRAs. When a mutual fund company, be it Fidelity, Vanguard, or T. Rowe Price, makes a trade to add to a position or subtract from one, it's inevitably a very large order. Trading a million shares of a stock, even a heavily traded one such as Apple or ExxonMobil, can move the market against the large seller or buyer. If other traders know that an order to buy a million shares is coming through the pipe, they will do whatever they can to get in front of it and buy up available shares. That way, they can repost their newly acquired shares for sale at a higher price knowing that they'll get sucked into the mutual fund order.

When a mutual fund has to pay more for its shares, it costs the owners of that fund—normal people saving for retirement—money. To combat traders prowling for big orders to abuse, mutual funds and other institutional traders hired hackers to develop algorithms that randomly disguise their large trades by breaking them up into smaller ones. The idea is a bit like stealth technology. Instead of appearing as a giant chunk of metal in the sky, the design and coverings on a stealth plane show radar hundreds of smaller objects, which could be cloud cover, birds, or other detritus. Dispatching sorties to intercept every bird pack across a continent is futile, which is why stealth worked. But the Russians eventually developed ways that could sometimes discern true packs of birds from eighty tons of American steel. The same thing happened on Wall Street. Traders who were at first defeated by algorithms disguising big institutional orders soon developed complex scheming algorithms of their own that could sniff the market for large trades that had been disguised.

Mutual funds and institutions dealing in giant blocks of stock reacted by building better algorithms of disguise—bots that issued decoy trades and would easily shift strategies if they suspected detection. The hunters, again, responded in kind, leading us to the current state of the stock market wherein algorithms are the gladiators and normal people, just like in the days of the Colosseum, stand by and watch. It's a fantastic game, although it has little to do with the base mission of the stock market: allowing growing companies an easy path to raising capital while giving the public a chance to invest in something that can build wealth in a meaningful way.

Peterffy thinks things have gone too far. He has already stepped back his company's participation in making markets. The man who was one of the most powerful forces behind the algorithms that have taken over Wall Street—and are now headed everywhere else—has second thoughts. "I only saw the good sides at the time," he says.

At the time, of course, Peterffy was building up one of the largest fortunes in America, so this might seem like an easy thing for him to say in hindsight. Peterffy thinks that in this age of light-speed trading, bids and offers on stocks should be held up for a minimum amount of time, still far less than a second, but enough to eliminate the head fakes, parries, and trickery that comprise the contemporary market and that lead us to clifflike falls and rocketlike spikes.

His ultimate fear is that a rogue series of algorithms sparks a string of colossal losses that their owners can't cover. Because some high-speed trading algorithms are able to trade on margin with leverage, it's conceivable that a series of bad trades, all conducted in seconds, could lead to a liquidity crisis, bankrupting a trader's broker and the clients he trades for. Such incidents have nearly happened before. In late 2009, Chicago's Infinium Capital Management, one of the more secretive and powerful trading houses in the United States, twice lost control of an algorithm that began selling S&P 500 futures as fast as it could, dropping the market. It happened again to Infinium in February 2010, when a new algorithm meant to capture small profits on crude oil trades tore up the commodities market as it traded wildly, losing more than $1 million in three seconds. The firm was fined $850,000 by the Chicago

Mercantile Exchange for "failing to diligently supervise its systems." The CME said that by unleashing such a flawed algorithm "to operate in a live trading environment, Infinium committed an act detrimental to the welfare of the exchange."[8] It was reported that Infinium, rather than testing the algorithm for six to eight weeks, as was its standard, set it loose after only two hours of checks.[9]

This virtual world of warring algorithms is what has become of Wall Street and our money. It doesn't stop there. The bot saga on Wall Street offers clues to what's in store for much of our future world.

2

A BRIEF HISTORY OF MAN AND ALGORITHMS

THE HIGH MATH BEHIND THE most brilliant algorithms is currently enjoying a renaissance. There have never been more people who understand it, and there have never been more people working to spread their understanding through discussion and investigation. To see this in action, all one needs to do is head to Y Combinator's Hacker News message board,[1] which has grown into one of the more influential Web sites in the world.

Here, hackers, math folk, entrepreneurs, Wall Street programmers, and people who are generally in tune with the waves of the Web come to discuss almost everything. Much of the discussion is about programming, startups, and Silicon Valley. But there's invariably a post or two on the front page discussing Gaussian functions, Boolean logic, or some other branch of mathematics that makes this increasingly algorithm-centric world possible.

The takeover of our world by algorithms, whether in an accounting office or at a customer service desk, has been looming ever since the math enabling such things was uncovered. We praise those who conquer our markets and rework society with algorithms and bots—but they were able to do so only because of theories constructed 250 years ago.

WHERE DID ALGORITHMS COME FROM?

Humans have been devising, altering, and sharing algorithms for thousands of years, long before the term itself ever surfaced. Algorithms needn't involve graduate school math or even math at all. The Babylonians employed algorithms in matters of law; ancient teachers of Latin checked grammar using algorithms; doctors have long relied on algorithms to assign prognoses; and countless numbers of men from all corners of the earth have used them in an attempt to predict the future.[2]

At its core, an algorithm is a set of instructions to be carried out perfunctorily to achieve an ideal result. Information goes into a given algorithm, answers come out. In their freshman-year programming classes, many college engineers design a simple algorithm to flawlessly play the game of tic-tac-toe.[3] In their program, the opposing, or human, player's move forms the input. With that information, the algorithm produces an output in the way of its own moves. A student expecting an A on such a problem will produce an algorithm that never loses a game (but often plays to a draw).

The algorithms used by a high-frequency trader or a speech recognition program work the same way. They're fed inputs—perhaps the movements of different stock indices, currency rate fluctuations, and oil prices—with which they produce an output: say, buy GE stock. Algorithmic trading is nothing more than relying on an algorithm for the answers of when and what to buy and sell. Building an algorithm with many variables is more difficult than building one to play tic-tac-toe, but the idea is identical.

The word *algorithm* comes from Abu Abdullah Muhammad ibn Musa Al-Khwarizmi, a Persian mathematician from the ninth century who produced the first known book of algebra, *Al-Kitab al-Mukhtasar fi Hisab al-Jabr wa l-Muqabala* (The Compendious Book on Calculation by Completion and Balancing). Algebra's name comes straight from the *al-Jabr* in the book's title. As scholars disseminated Al-Khwarizmi's work in Latin during the Middle Ages, the translation of his name—"algorism"—came to describe any method of systematic or automatic calculation.[4]

The first algorithm recorded and later found by modern civilization comes from Shuruppak, near modern Baghdad. The Sumerians, who ruled their piece of the Euphrates Valley for fifteen hundred years, left behind clay tablets dating from roughly 2500 BC that illustrate a repeatable method for equally dividing a grain harvest between a varying number of men. The method described utilized small measuring tools; it was useful because vendors of that time didn't have scales large enough to weigh thousands of pounds of food at once. The tablets carrying this algorithm, depicted in symbols, now sit in the Istanbul Museum.[5]

Some algorithms developed thousands of years ago still play very real roles—albeit ones their creators couldn't have imagined—in the computerized world of today. Many Web sites, wireless routers, and other places where passwords and user names must be encrypted employ an algorithm that was conceived more than two thousand years ago by Euclid of Alexandria, a Greek mathematician.

The Euclidean algorithm, as it's often referred to by math scholars, dots programming within dozens of modern industries and can be used to derive most, if not all, of modern music's rhythm patterns.[6] Around 300 BC, Euclid wrote his *Elements,* a work that would form the backbone of geometry texts for the next twenty-three hundred years. In *Elements,* Euclid included an algorithm that finds the largest divisor of two different numbers. Anybody with fifth-grade math skills can use it to quickly determine that 17 is the largest common divisor of 1,785 and 374.[7]

THE GOLDEN MEAN

People who work in biology, botany, astronomy, and even architecture should all be familiar with a concept developed when Europe was first adopting modern numerals and the decimal system in the twelfth century.[8] Mathematicians in particular began toying with the curious ratio of 1.618—known as the golden mean—a number often exhibited in nature, such as in the fractal geometry of ferns, the atomic structure of DNA, and the orbital patterns of galaxies.[9] Architects with keen senses of proportion and scale gravitate toward the ratio as well. Great buildings that look "right" often have spacings and patterns based on a ratio of 1.618. Le Corbusier, the influential twentieth-century Swiss-French architect, explicitly used the ratio in many of his buildings, such as Villa Stein in Garches near Paris, beginning in 1929. More recently, graphic designers have spotted the ratio in many of Apple's products, including its logo.[10]

Leonardo Fibonacci, the man behind the golden mean, was most responsible for Europe's adoption of modern numbers and is considered by many historians to be the most important mathematician of the Middle Ages. Before the rise in central Italy of what would come to resemble modern banking, Leonardo of Pisa traveled widely within the Mediterranean basin and even to the Orient; his father was a Pisan customs official based in what is now Bejaïa, Algeria.[11] Fibonacci, as he's known to most, saw the utility of Arabic-Indian numbering systems compared with clunky Roman numerals, especially when it came to mathematical computations.

Fibonacci published *Liber Abaci* (The Book of Calculation) in 1202. Within the book, the young math whiz explained how decimals relate to fractions and how they can be wielded to make bookkeeping simple and solve real-life problems. Fibonacci dispatched conundrums common for the time; in his examples, he includes methods for dividing staples such as peppers, animal skins, and cheese. While citing Al-

Khwarizmi, Fibonacci provided Western civilization with a series of algorithms that would be heavily used for centuries, such as those for calculating the present value of future money streams and for interest payment systems that resemble our modern mortgages.[12]

Whereas most ancient math scholars' significance gets left to dusty history tomes, Fibonacci's name remains well known on Wall Street as well as in popular culture. Novelist Dan Brown put a spotlight on Fibonacci in *The Da Vinci Code,* a book that has sold nearly one hundred million copies. Brown keyed on the curiosity of the so-called Fibonacci sequence, included in *Liber Abaci,* in which each number is the sum of the two digits preceding it: 1, 1, 2, 3, 5, 8, 13, 21, 34, etc. As the sequence progresses, the ratio of its numbers and their immediate predecessor converges toward the golden mean of 1.618.

On Wall Street the number has also kindled plenty of belief in its powers, as some quirky traders have ventured billions of dollars on algorithms based on the golden mean or the so-called Fibonacci numbers. More than a hundred books have been published touting the power of the golden mean and its presence within marketplaces as diverse as commodities, stocks, and foreign exchange. And that's just in the last twenty years. There's little evidence that any of this is true, but Wall Street is nothing if not a paradox where cockamamie theories and rigid logic happily coexist in equal abundance.

THE GODFATHER OF THE MODERN ALGORITHM

Gottfried Leibniz, like Isaac Newton, his contemporary, was a polymath. His knowledge and curiosity spanned the European continent and most of its interesting subjects. On philosophy, Leibniz said, there are two simple absolutes: God and nothingness.[13] From these two, all other things come. How fitting, then, that Leibniz conceived of a calculating language defined by two and only two figures: 0 and 1.

Leibniz developed this system to express numbers and all operations of arithmetic—addition, subtraction, multiplication, and division—in the binary language of 1's and 0's. The mathematician defined the language in his 1703 paper "Explanation of Binary Arithmetic."

Leibniz was born in Leipzig in 1646 on a street that now bears his name. An overachiever from the start, he began university at fifteen and had his doctorate by the time he was twenty. Known for being gregarious, he chatted up his physician on the subject of alchemy as he lay dying in bed at the age of seventy.[14] The large shadow of Leibniz in this story traces to his convictions about the breadth with which his binary system could be applied. He thought far beyond the erudite world of math theorems and the Newtonian wars of calculus theory.

All physical changes have causes, Leibniz said. Even people, to some degree, have their paths set by the outside forces acting upon them, a fact that game theory, developed long after Leibniz's time, exploits. Leibniz therefore believed that the future of most things could be predicted by examining their causal connections.[15] This is what much of the modern Wall Street titans have realized better than anybody else. Had the man been born later, much later, he could undoubtedly have raised a large and successful Wall Street tent.

Before anyone else, Leibniz conceived of something reaching toward artificial intelligence. The mathematician stipulated that cognitive thought and logic could be reduced to a series of binary expressions. The more complicated the thought, the more so-called simple concepts are necessary to describe it. Complicated algorithms are, in turn, a large series of simple algorithms. Logic, Leibniz said, could be relentlessly boiled down to its skeleton, like a string of simple two-way railroad switches comprising a dizzying and complicated national rail network. If logic could be factored to a string of binary decisions, even if such sequences stretched for miles, then why couldn't it be executed by something other than a human? Leibniz's dream of reducing all logical thought to mechanical operations began with a machine he designed himself.[16]

After hearing of an adding machine built by Blaise Pascal, Leibniz set out to best him. His machine would perform addition and subtraction more smoothly and could also solve multiplication and division problems, something Pascal's machine couldn't do at all. Putting the design down on paper, Leibniz hired a Paris clockmaker to make the machine reality in 1674.[17]

Leibniz crossed the English Channel to demonstrate the machine's efficacy to the Royal Society in London, the arbiters of the day's intellectual prestige. But the machine failed during his demonstration and Leibniz was reduced to carrying division remainders by hand.[18] He seemed to lose interest in the machine after the exhibition, but his design remained the dominant one for two hundred years, with subsequent generations of calculators built on its plan, which had been recorded. The machine itself was lost to history for more than two centuries. Leibniz apparently stored the brass cylinder in the attic of a building at the University of Göttingen, where it remained in a low-slung corner until a maintenance crew, sent upstairs to fix a roof leak, found it in 1879.

Beyond calculating machines, Leibniz believed that by breaking logic and thought down to arithmetical calculations, he would find a calculus ratiocinator, a kind of algorithm that would resolve arguments. He imagined that disagreements wouldn't need to be settled by one opponent shouting down the other. Instead, two adversaries would proceed to a table with pencil and paper in hand, ready to calculate who, according to logic, was right.[19] If U.S. politics gives us any indication, Leibniz never found his sacred calculus ratiocinator.

Controversy marred the end of Leibniz's life. Although it's now universally accepted that he and Newton discovered calculus in parallel, at the time, Leibniz, who often corresponded with Newton by letter, was accused of pilfering the Englishman's discovery for himself. And while it may have been hard to share the stage with Newton, the most prolific mathematician and physicist who ever lived, Leibniz did plenty to distinguish himself. As for calculus, it was he, not Newton, who developed elegant notations for integrals and derivatives that all calculus students still learn today.

Leibniz's development of the symbols and theory behind calculus, which allows for the precise study and modeling of change, gave mathematicians a weapon with which to create algorithms powerful enough to build semiconductors, connect us through radio,[20] and launch satellites into orbit with a laser's precision. Calculus and algorithms have intertwined histories, meanings, and power. The relationship is well summed up by mathematician David Berlinski in his book *Infinite Ascent:*

> There now occurs a reverberating sonic boom! in the history of thought. Before the discovery of calculus, mathematics had been a discipline of great interest; afterward, it became a discipline of great power. Only the advent of the [computer] algorithm in the twentieth century has represented a mathematical idea of comparable influence. The calculus and the algorithm are the two leading ideas of Western science.[21]

No one who lived before the 1960s has ever been commonly referred to as a hacker, or a quant. But perhaps such a moniker should be extended to this German who lived during the latter days of Europe's Renaissance. Leibniz moved algorithmic science forward in three distinct ways. He was one of the founding instigators of calculus, and, just as important to this discussion, he introduced the method of building algorithms to express complicated solutions out of a series of simple binary blocks.

Leibniz's third important contribution to the power of algorithms lay in the link he sought between the simplest fragments of language and the human emotions they reveal. Leibniz thought language and how humans use it should be studied in a rigorous academic way. He figured that if something as complicated as man's existence could be broken down into two absolutes, be they God and nothingness or 1 and

0, why couldn't language be deconstructed in a manner in which paragraphs, sentences, clauses, and words could be sifted for more meaning? The philosopher-mathematician speculated that humans uttered words and phrases that fit their individual emotions and perceptions.[22]

By knowing what a person says, we can know who they are. By knowing who somebody is, it becomes easier to predict what exactly they'll do in the future. At the heart of all of these algorithm-enabled revolutions on Wall Street and elsewhere, there exists one persistent goal: prediction—to be more exact, prediction of what other humans will do. That's how money gets made. Leibniz's hunch—that individual humans are programmed to behave in certain predictable manners—was more accurate than many knew, and it's a fact that powers many Wall Street algorithms today. Later in the book, we'll see how this human predictive science that Leibniz first speculated on was developed at NASA and how it's now spreading into the rest of our lives.

Leibniz never saw his binary creation reach the heights he envisioned for it. But beginning in the 1930s, electronic circuitry capable of executing arithmetic problems began to manifest in the United States, Germany, and France. It utilized Leibniz's binary numbering system, which had been waiting 250 years for material science to catch up to its simple brilliance.

The binary system enables all of today's computer programming languages, which are nothing more than vehicles for easily writing algorithms. It also powers the chips and circuits of our computers that run those algorithms.

GAUSS: MAKING THE LOGIC BEHIND ALGORITHMS POSSIBLE

Algorithms only work when we know the exact factors that need to be examined or manipulated. An algorithm that determines the fair worth of a stock option would be impossible to construct without knowing that this price is wholly dependent on the volatility of that stock's historical prices, interest rates, and the strike price of the option. If Peterffy or Black and Scholes didn't know that these were the most important factors affecting options' prices, then creating an algorithm to effectively trade and defeat humans would be unfeasible. The same goes for an algorithm that knows what you're really thinking by examining the words you speak. This algorithm has to be imbued with the logic of what our speech patterns and sentence structure actually mean.

Sometimes figuring out which variables matter most in these relationships is as hard as constructing the algorithm that will automatically put these findings to use. In a model such as the stock market or, say, the number of words we utter in a day, thousands of things can be categorized and searched for underlying meanings. But most of the data is just noise; it's meaningless. To work properly, predictive algorithms need instructions that allow them to ignore the noise and zero in on the factors that matter. Teasing out these important factors, be they what truly affects options prices or what makes a song catchy to our ears, is done by analyzing heaps of data. The most common method of sifting data and dusting off the most interesting and unintuitive relationships is called regression analysis, a technique that allows a physicist, a statistician, or an engineer to make accurate predictions based on data from the past. The development of methods to make models fit existing data owes at least part of its existence to the English royal family.

The royals weren't so clever that they figured this out for themselves. But King George IV was smart enough to commission Carl Friedrich Gauss in 1817 to survey the Kingdom of Hanover.[23] Gauss, already an accomplished mathematician, found the surveying tools of the day to be inaccurate, so he invented a new one, called a heliotrope, which used a mirror to reflect sunlight across large distances. Even with this new instrument, Gauss knew that an attempt to measure a land so large would be prone to error. He had roughly developed a method for dealing with this sort of problem twenty years previously when he was eighteen. It's known as the least squares method.

The least squares method allows for predictive model building based on observed results. To find the best model according to the data in hand, Gauss developed equations that minimize the squares of the differences between measured values, which could be stock prices of the past, and values provided by a predictive model. The model—which can be anything from a quadratic equation to a multitiered algorithm—is adjusted up and down until the point of least squares is found. The least squares method forms the backbone of modern statistics and algorithmic model building. When building an algorithm to, say, predict future stock prices, testing it means running the algorithm through loops of data representing days and prices that have already taken place. A function, or curve, is derived from the past data that the operator hopes will be predictive of future data. When quants, engineers, or physicists talk about data fitting or regression analysis, they're likely wielding Gauss's innovation.[24]

In his research on the distribution of measurement errors, Gauss would also conclude that deviations (the errors) had what's called a normal (now also known as Gaussian) distribution. Such a distribution, when plotted as a curve, resembles a bell. Gauss and others of the era observed that most measurable things, even those in nature such as the heights or head sizes of people, maintain the same distribution.[25] There exist a large bunch of data points near the mean value, in the middle of the bell, and fewer and fewer data points as the bell spreads out, away from the mean. Similar distributions are seen on student test scores, the size and height of people, and a multitude of things in nature.

The far edges of the bell curve are referred to as the tails. In a Gaussian distribution, those tails taper off toward zero. In real-life situations that can be affected by the irrationality of humans, such as financial markets, the tails aren't so thin. When human psyches and behavior are involved, there is a higher chance of outlier events, producing a distribution with so-called fat tails, something that is not Gaussian. Wall Street fortunes have been made betting on Gaussian distributions—and just as many have been lost on algorithms that embrace Gaussian outcomes but don't account for fat tails.

It's easier to write algorithms to fit normal distributions. And despite history showing us repeatedly that human behavior is anything but normal, some hackers choose to account for only normal distribution. Using this assumption can make money 100 out of 100 days. But it's day 101, the Black Monday of 1987, the Russian debt default of 1998, the Flash Crash of 2010, that can ruin those banking on algorithms designed purely around Gaussian distributions. Even Gauss, more than two hundred years ago, warned that errors of any magnitude are possible within a normal distribution.[26]

The introduction of normal distributions changed humankind and ushered in the modern field of statistics, which allows for the easy purchase of things like life insurance, the building of better bridges, and even, though not as important, betting on basketball games.

Despite much success early in his life, Gauss stayed relatively humble, avoiding controversy and only publishing work he considered utterly irrefutable. He considered Euler and Newton to be the greatest mathematicians.[27] But most experts would include Gauss himself within their company. Many scholars say Archimedes, Newton, and Gauss comprise math's immortal triumvirate. There exist dozens of tales of Gauss's early talent for mathematics, some of which recount his amazing arithmetic skills at age three.[28] One story tells of him in school at seven years old when his teacher assigned his class the task of adding up all of the numbers from 1 to 100. A few seconds later, Gauss piped up with the answer, apparently having deduced in his head an algorithm used by the Pythagoreans as a password for their secret society: [29]

$$\frac{1}{2}n(n + 1) = S$$

The variable n denotes the last number in the sequence, in this case 100, and S equals the sum. The young Gauss likely noticed that 1 + 100 totaled 101, as did 99 + 2 and 97 + 3, which meant there were 50 sets of 101, or 5 × 101 = 5,050.

Born in 1777, Gauss came from a poorer family, a long way from the normally gilded class of leading intellectuals. When he was fourteen, his mathematical talents drew the attention of the Duke of Brunswick, who assigned him a stipend and, later, a spot at the University of Göttingen.[30] His mind produced ideas so rapidly that he sometimes had trouble writing all of them down.[31]

Gauss earned his PhD at age twenty-one with a dissertation that offered the first complete proof of the fundamental theory of algebra, something Newton and Euler had each attempted but failed to finish.[32] Gauss's mind would lead him to discoveries in many corners of mathematics, including number theory, quadratic equations, and the general distribution pattern of prime numbers among the integers.

Some may associate Gauss's name with that of the Gaussian copula, an often demonized formula introduced to Wall Street in 2000 by David X. Li. In statistics, a copula is used to determine the behavioral relationship between two or more variables. Li was searching for a way to measure the risk of one mortgage defaulting if another, seemingly unrelated mortgage defaulted. This is called correlative risk. Because Li didn't have piles of historical default data on subprime mortgages (it simply didn't exist), he built his copula on data that did exist: historical prices of credit default swaps, which result in a payment to the owner of the swap if the underlying securities (mortgages in this case) go into default. But the CDS market was, as we now well know, egregiously mispriced by the humans who traded the swaps and set the prices. Nevertheless, Wall Street embraced Li's formula as stone-solid fact. The copula should have been one arrow in the quiver of analysts and rating agencies who examined and stamped their approval on mortgage-backed securities. Instead, it became the only arrow.

The resultant boom in collateralized debt obligations and the housing market bubble came straight from bankers' misuse of what should have been a harmless algorithm. Gaussian copulas are useful tools and are utilized in a number of fields, but the one thing they do not do is model dependence between extreme events, something humans excel at precipitating.[33]

PASCAL, BERNOULLI, AND THE DICE GAME THAT CHANGED THE WORLD

Much of modern finance, from annuities to insurance to algorithmic trading, has roots in probability theory—as do myriad other businesses from casinos to skyscraper construction to airplane manufacturing. The most promising movement in modern medicine bases much of its testing and diagnostic methodologies on probability. Political parties make decisions on candidates and the direction of their ethos not based on pragmatism or facts, but on probability theory. Football coaches consult pocket cards instructing them on what decision to make based on the probabilities of other game events happening in the time left on the clock.

All of these modern manifestations of probability theory can be traced to a letter sent between two Frenchmen in 1654 when Blaise Pascal wrote a note to Pierre Fermat contemplating how one might break up the pot of a game of chance that has yet to be finished. One player, for instance, has won five rounds and the other has won two—how much does each take home from the communal pot? Pascal's three-thousand-word essay established a method for the exact calculation of events' likelihoods. The answer to the gaming question is to calculate the likelihood of each of the remaining players winning the pot and split it according to those percentages.[34]

Within five years of that envelope crossing France from one famous mathematician to another, the world began to change accordingly. Being able to better calculate what might happen in the future gave rise to burgeoning industries and allowed people to manage their lives and businesses with far more certainty. Pascal's letter to Fermat is what allows farmers to disconnect market risk from their grain crop; it's why, just one hundred years later, mortality tables were being issued with precision by Europe's financial sector. Pascal's letter is one of the chief reasons why London ascended to become the world's most consequential city for at least a century as its merchants insured much of the world's largest seagoing fleet, encouraging businesses to strike out for new lands without fixating on the enormous risks such ventures entailed.

Building on Pascal's work, Jacob Bernoulli, like Christiaan Huygens a few years before him, became transfixed with putting rigid probabilities on games of chance involving things such as dice and cards.[35] It was through his study of these games that Bernoulli developed what is now known as the law of large numbers.[36]

For processes with probabilities that can be calculated, such as the flipping of a coin (50 percent chance of heads, 50 percent chance of tails), Bernoulli's law said that as more iterations of the game are played (or coins are flipped), the average result will converge upon the precalculated probability. Turning heads in 70 percent of ten coin flips, for instance, isn't all that unlikely. Getting heads on 70 percent of a thousand coin flips, however, is virtually impossible.

This principle guides blackjack gamblers and poker players. It also guides many, if not most, of Wall Street's high-frequency traders, who keep their trade volumes high to ensure that their net hones close to a probable mean. Probability theory also guides modern baseball general managers, who look for players with talent whose regular stats, such as earned run and batting averages, have been skewed lower by a prolonged string of bad luck. Such misfortune might be revealed by more complex stat measurements that show how, in the case of a pitcher, perhaps, an outsized number of ground balls have turned into hits, or how, in the case of a batter, line drives haven't resulted in hits. There is also a new class of doctors who, in diagnosing patients, before all else ignore their biases and look to algorithms based on probabilities.

In the case of a stock trader, an algorithm that seeks to capture the spread between a stock's bid and offer prices can be thwarted when the market moves, which can leave the trader and his algorithm with only one side—the wrong side—of a trade. If an algorithm bought Microsoft at the bid of $50.00 and then failed to sell it within the same several seconds at the ask of $50.02 before the market went down, the trader will be taking a loss when he sells at the new ask of $49.99.

These algorithms are built to predict market direction. To be successful, the algorithm has to be right only 51 percent of the time. But a 51 percent probability could still leave a high-frequency trader open to big losses if he trades only ten times a day. There could be plenty of days where he loses on seven of ten trades. But high-frequency traders are called "high frequency" for a reason. They make tens of thousands of trades a day representing millions of shares. As their repetitions escalate, as Bernoulli pointed out, their number of profitable trades will converge toward the mean of 51 percent. This is why many of the biggest, most powerful algorithmic traders can go a year or longer without having a losing day.

Beyond Wall Street, the math mortals among us can thank Bernoulli for his work on defining formulas for calculating compound interest, one of the first ideas most people learn and retain when it comes to finance. Within his algorithm for continually compounding interest, Bernoulli uncovered the mathematical constant known as *e*. In correspondence with Leibniz, Bernoulli referred to the number as *b*, perhaps for his own last name, but it became known to the world as *e* after a man named Leonhard Euler referred to it as such.

GIVING VISUAL SHAPE TO ALGORITHMS

In 1791, the famous Austrian composer Joseph Haydn attended a grandly staged version of George Frideric Handel's *Messiah* at Westminster Abbey in London. At the close of the performance, put on by a thousand choral and orchestra members, Haydn wept. Through his tears, he declared of Handel, his contemporary, "He is the master of us all."[37]

At about the same time, Pierre-Simon Laplace, the French mathematician and one of the thought giants who developed the field of statistics, was exclaiming the same thing, but not of the *Messiah*'s composer.

The man Laplace proclaimed the "master of us all" was Leonhard Euler.[38]

Euler was another product of the University of Basel, a world-altering cluster of intelligence. Pope Pius II founded the university, the oldest in Switzerland, in 1460, and for centuries it drew in high intellects such as Erasmus of Rotterdam, the Bernoullis, the Eulers, Jacob Burckhardt, Friedrich Nietzsche, and Carl Jung. Born in 1707, Euler was schooled for a time by Johann Bernoulli, Jacob's brother, who was at the time likely the world's greatest working mathematician.[39]

While at university, Euler spent his Saturday afternoons talking with the younger Bernoulli brother about mathematics and philosophy. "He kindly explained to me everything I could not understand," Euler remembered.[40] By the time his studies were finished, there wasn't much left needing explanation. Euler then began perhaps the most prolific stretch of mathematical publication ever undertaken by one man.

As a new professor at the Saint Petersburg Academy of Sciences in Russia, Euler tackled a problem known as the Seven Bridges of Königsberg, which had bedeviled many people in this part of the continent. Königsberg, now called Kaliningrad and situated in a small exclave of Russia wedged between Poland and Lithuania, was in Euler's time a part of Prussia. The city was divided into several landmasses, including two islands, by branches of the Pregel River, and these were connected by a series of seven bridges. A popular puzzle for those strolling the city would be to try to find a route through the town that crossed each bridge but only once.[41]

To illustrate the problem, Euler drew a series of points, called nodes, connected by lines, called edges, representing the landmasses and the bridges. He noted that the distances and the shape of the lines could be altered in any fashion and that the nodes could be moved around, so long as all of the lines (bridges) remained intact. What Euler created is now known in mathematics as a graph, and it proved that crossing Königsberg using every bridge once and only once was in fact impossible. In solving this problem, Euler developed graph theory.

The graph theory that Euler developed is separate from the graphs we are used to viewing of, say, the stock market or a sales report. Euler's graphs are treelike diagrams that can symbolize networks in nature, circuitry on a microchip, or relationships between different humans within a city. Writing algorithms specifically for graph theory represents one of the exciting new leafs of modern computer science, allowing biologists to make connections between DNA strands and physical traits, professors to decode the music of the Beatles, the CIA to connect terrorists across the world, and Wall Street observers to find relationships between seemingly disparate things. Graph theory is especially useful to those examining Facebook networks, where the most influential members can be determined by analyzing which nodes in the graph (people) draw the most edges (connections) and which edges are most active in terms of drawing eyeballs and comments. Engineers who interview for jobs at Facebook are often quizzed on graph theory.

Euler lived for seventy-six years, but his work continued to be published for nearly a century after his death. He authored 886 books, articles, textbooks, and technical manuals, representing a full third of all European mathematical publishing during his lifetime. He blazed so many new paths in mathematics that, in an effort to avoid naming so much after one man, many theorems and equations carry the name of the first person who discovered or applied them after Euler.[42]

Some people might remember from school what's known as Euler's formula: $V - E + F = 2$. The equation, upon which countless algorithms are written, describes three-dimensional shapes with V equaling the number of vertices (corners where lines intersect), E being the number of edges for the shape, and F equaling the number of faces of the object. A cube, for instance, has six sides or faces, twelve edges, and eight vertices: $8 - 12 + 6 = 2$.

This equation took Euler far beyond elementary and rigid shapes like cubes, pyramids, cones, and spheres. Euler's formula would later help explain the geometry of carbon molecules, seemingly random weather systems, optics, magnetism, and hydrodynamics. Through the theory behind his namesake formula, Euler began pondering nonrigid shapes, the study of which is known as topology. Topology is part of chaos theory, which in the last two decades has yielded fortunes to mathematicians who came to Wall Street and built algorithms around it.

BOOLEAN LOGIC MACHINES

All of the math geniuses mentioned here contemplated human thought, its origins, its limits, and its methods. Leibniz was first to theorize that human thought could be broken down to its basest of cogs, represented by a series of binary decisions. These binary choices, Leibniz said, could then be stacked on top of one another, as high as needed, to form thoughts—or algorithms—growing more complex.

If Leibniz took the first step toward building the machines that now rule our lives, it was George Boole who seized Leibniz's momentum, which was by his time nearly two hundred years old. It's Boole's calculating system and innovative form of algebra that allows the Web to work, from the images people post on Facebook to the text they post on their blogs. The complex algorithms that have come to rule our lives would be nothing if they couldn't make humanlike jumps in their logic. For example, Google will show a Gmail user his in-box only *if* he correctly types his password *and if* he, if prompted, correctly deciphers that squiggly word that's hard to read. Google will *not* show you an e-mail that is marked as spam because the same e-mail went to ten million Gmail boxes *and* it's *not* from a trusted mass sender such as Groupon *or* somebody with whom each one of these people has already held an e-mail conversation. It's these modifiers, the *if*s, *and*s, *or*s, *not*s, that make computer science and algorithms tick.

The roots of this revelation struck a seventeen-year-old Boole while he strolled across a meadow. From out of nowhere, he formed the idea that algebraic symbols of some sort could be used to define a language of logic, a language of thought—a way to break down the inner workings of human rationalization. The idea proved so revolutionary to the world and so important to Boole's life that he began to ponder the power of an unengaged mind. Science has well proven that the human subconscious is perhaps man's strongest tool, but nobody theorized on the fact before Boole. He referred to it as "the unconscious."[43]

Boole was born in 1815 in Lincoln, England. His father, John, was a shoemaker, but he possessed a bevy of erudite hobbies, including mathematics, astronomy, and the construction of optical instruments. John Boole couldn't afford to send his son to college, so George taught himself. His autodidactic methods worked so well, in fact, that he could read and write in four languages before he reached his teen years, by which time he was teaching in local schools.[44]

In his twenties, Boole began to attack problems of advanced math; journals began publishing his work in 1841. Four years later, he was named the founding professor of mathematics at Queen's College in Cork, Ireland (now called University College Cork). In 1854, Boole harvested the seeds that had started to bloom in that meadow when he was seventeen, publishing *An Investigation of the Laws of Thought, on Which Are Founded the Mathematical Theories of Logic and Probabilities.* With this effort, he chased something Leibniz had thought of long before. But it was Boole, not Leibniz, who managed to corral the idea into a medium that could be used.

Boole opened his book with this:

> The design of the following treatise is to investigate the fundamental laws of those operations of the mind by which reasoning is performed.

The Englishman sought to define a language of thought that could break human deduction into a series of mathematical expressions easily written onto paper, creating symbols that represented the gates within human thought: *if, and, or, not,* which could be used with normal arithmetic operations such as multiplication and division. Anybody who has written just an elementary chunk of computer code is familiar with these modifiers. Boolean algebra gives computer circuitry its guts. Without it, the execution of endlessly complex algorithms—the kinds that now change the world every day—would be impossible.

Boole's ideas did not set the world on fire after he published. Few Britons, including the country's mathematicians, were familiar with logic theory. And there was no obvious way to apply what's now called Boolean algebra without machines or computers capable of reading algorithms.

That's not to say nobody tried. Perhaps the most important was Ada Lovelace, who, aside from being a female math scholar in a time when few women were allowed to study, is often recognized as the first hacker. In 1842, while documenting Charles Babbage's Analytical Engine, a mechanical computing machine that Babbage never completed, Lovelace devised several different inputs that, theoretically, would make the machine perform certain calculations and tasks. In doing so, Lovelace had composed the first algorithm meant for a machine. Lovelace's work did not use Boolean symbols, so it was left searching for a way to express the processes of human thought. She wrote that she wanted the device to "feel about" to discover the correct decision or path.[45] In 1979, well over a century after Lovelace's death, the U.S. Department of Defense named its new computer language Ada, after the original hacker.[46]

Despite the attempts at mechanical computers by Babbage, Lovelace, and others, Boole's epiphany, while respected as a mathematical theorem in the early 1900s, hadn't caught on with the engineers and physicists who started building the electronic age.[47] But as technology progressed and scientists attempted to build more and more advanced circuitry, engineers found themselves needing something—a mathematical tool—that would help them tame and fully utilize the sophisticated electronics they were building.

In the late 1930s, an MIT graduate student named Claude Shannon combined the binary calculating and number system of Leibniz with

Boolean operators such as *and, or, not, not-or,* and *if,* among others. Shannon found that all of these expressions could be built into electronic circuits—and they could also solve nearly any math problem, store data, and edit all kinds of information from images to text. Shannon, who achieved much in his long life, is known as the father of the information age.[48]

And here was the true birth of computer circuitry and language, and with it the modern algorithm and the beginnings of machines that can mimic man.

3

THE BOT TOP 40

IN 2004, A MAJOR MOVIE studio allowed an algorithm to scan nine scripts of unreleased movies. The results of the analysis, run by a new company named Epagogix, were then tucked away. The movies all eventually hit the screen, and when the last one was out of theaters, the movie studio went back to take a look at what the algorithm, which was supposed to tell them how much money each film would gross at the box office, had predicted. In three of the nine cases, the algorithm missed by a wide margin. The other six forecasts, however, were bizarrely accurate. On one movie that the studio expected $100 million or more on, the total gross was $40 million, a huge disappointment.[1] The algorithm predicted $49 million. Another prediction was within $1.2 million. Epagogix was suddenly on its way to becoming an indispensable tool for studios to use in analyzing scripts—especially ones that may be attached to big budgets—before the movie gets made. Epagogix was conceived and built by two movie lovers, one of them a lawyer and the other from Wall Street's favorite of disciplines: risk management. The point is to minimize the risk of producing a stinker like

Disney did in 2012 when *John Carter* lost the studio nearly $200 million.

What the algorithm actually scans is a report produced by people who read the script and grade it on hundreds of different factors, from its setting to the kind of heroes who lead the film to the moral dilemmas within the plot to the supporting characters, the ending, the love stories, and so on. No need for focus groups, boardroom arguments, or executive clashes over whether to buy a new script; just grade it out and give it to the algorithm. But this algorithm, as genius as it is, still needs people to evaluate the words, the story, the plot, the characters, all of it. What if there were an algorithm that didn't need people for input? What if there were algorithms that could create the script itself?

Those who imagine art, innovations, words, novel strategies for companies, products that change the world—their jobs have always been considered outside the reach of algorithms. These professions and their perches hold esteem, good salaries, and a sense, for those who occupy them, of freedom and mobility within the workforce. Some call these people the creative class, some call them post–college educated, and some just call them smart people.

Smart people assume that this creeping revolution of bot workers can't touch them. The notion is that algorithms can't innovate, that a bot can't create. We're now learning, however, that these are dangerous assumptions.

Algorithms can be taught to evaluate the quality and originality of others' creations. They can generate their own creations too. One of the more unlikely areas for an algorithm invasion is music. To many, music reflects the beating creativity of the human soul. We can have difficulty describing how exactly music affects us, changes our mood, and alters our consciousness. Those who make music can barely explain how inspiration strikes. It's a process that's been chalked up as one of those indefinable sparks within the human subconscious—a mixture of right brainpower and long stretches of dedicated thinking on the matter, some conscious, some not. Creativity is thought of as something so incorporeal that it can hardly be taught, let alone left for a machine to carry out.

But there now exist algorithms, including one with a human name—Annie—that can produce music as daring and original as the works of masters like Brahms, Bach, and Mozart, and as popular and catchy as the tunes played inside a big-box store.

YOU HAVE A 41 PERCENT CHANCE OF BEING LADY GAGA

As a musician and writer, Ben Novak drove the car he could afford in 2004: a 1993 Nissan Bluebird. The vehicle propelled him around his hometown of Auckland, New Zealand, just fine. Novak's main complaint about the car concerned its radio, which could capture only two FM stations out of the dozens broadcasting in the city. As somebody who spent every spare minute imbibing or playing music, Novak found this no minor aggravation. But he didn't have money for a new car, so he left the dial fixed on the BBC, his only acceptable option.

Being stuck on the BBC had its benefits. Novak was well armed for cocktail conversation on current events, and he could always crack off a new piece of intellectual fodder when chitchat grew stale. More important, he didn't miss a short BBC report on technology developed in Spain that, the person being profiled claimed, could predict which songs would turn into pop hits.

"I'm driving down the road listening to this and I think, 'That's interesting,'" Novak remembers. "I could have just kept driving and forgot about it, frankly, but I was getting off at the next exit."

Novak hit his exit, drove to his house, and sat down in front of his computer. He brought up the Web site belonging to what was then called Polyphonic HMI. For fifty dollars, its algorithm would analyze any music file Novak uploaded. Potential hits earned high scores, duds got low ones.

"I mean, for fifty dollars—it was such a small amount of money when you think about what it could mean in the long run," Novak says. "So I did it."

Novak had written a song a few years before called "Turn Your Car Around" that he believed held significant potential. He uploaded the song and sat at his screen, waiting for a result.

Novak began writing songs when he was twelve and started playing the guitar in his late teens. He'd become so dedicated to songwriting by the time he was in his midtwenties that he'd forgone much of a social life so he could stay home on nights and weekends working on his music. He spent $3,000 on computers and midgrade recording equipment to outfit a spare bedroom as his music lair. He lined the closet with old comforters and positioned his best microphone in the center. This was the sound booth where he recorded "Turn Your Car Around."

Finally, the Web site whirred to life with an answer for Novak. The algorithm behind the site used a number scale to rate songs. Anything more than 6.5 had decent hit potential. Anything past 7 had a hook made for the pop charts. Novak's song scored a 7.57—as high as the algorithm had scored many of the biggest rock hits of all time, such as "Born to be Wild" by Steppenwolf and "Peaceful Easy Feeling" by the Eagles.

"I was really happy, obviously," says Novak. "But it wasn't really clear what came next."

In Spain, where Polyphonic HMI was based, the computer engineers who maintained the algorithm took note of the song for its high score. They pulled it down off the server and played it in the office.

"There was clearly something there," says Mike McCready, who ran the company. "Our guys played it over and over again." A musician himself, McCready called some recording label contacts in Europe and got the song circulating.

About two weeks after he submitted his song on McCready's Web site, Novak's phone rang. It was a representative for Ash Howes, a music producer in the United Kingdom with a few dozen hits in his pocket.

He had a young British pop star, Lee Ryan, who needed more tracks to fill out his album. Howe thought Novak's song would fit in well. In fact, he thought it could be a single.

Novak quickly agreed to a favorable deal: he would get 50 percent of all royalties when the song was played on the radio, on TV, or in an advertisement. Novak's song not only went on Ryan's album but was also designated the CD's first single. The song debuted at No. 12 on the English pop charts, and for two months in a row it was the most played song in the UK. It fared just as well on the European continent, where radio played it extensively and it cracked into a few advertisements. In Italy, the song ascended to No. 2.

The success changed Novak's life. He took some of his royalty proceeds and paid off a series of rental properties he had invested in. He bought better recording equipment and updated his instrument collection. There's one thing he didn't do, however, even by the time of this writing: replace that old Nissan. "It still drives just fine—and I'll never turn the BBC off at this point," he says.

Novak isn't shy about crediting an algorithm. The music world is one in which a hair's width of luck can make an artist or keep them from being discovered. Algorithms that sniff out talent can change that. "This whole music thing is just a huge gamble for anybody who goes into it," he says. "This program, this Web site—it aligned the planets for me."

The algorithm that changed Novak's life was devised by a group of engineers in Spain headed by McCready, an American who took an odd path to becoming an authority on the technology that's changing the future of music.

McCready grew up in rural Nebraska, a place that, he says, you either leave forever or stay forever. In 1986, as a high school senior, McCready was presented with his first chance to leave, albeit just temporarily in this case. He participated in an exchange program in which he left Nebraska for nine months to board with a family in a small village outside of Barcelona. When he returned, he headed to college at Creighton University in Omaha, where he majored in psychology. As he wrapped up his university years, McCready felt he had two choices. "I could go back to rural Nebraska or go back to Barcelona," he says.

McCready returned to Spain. Getting a job there, however, proved as difficult as it might have been in a Nebraska farm town. The Spanish government didn't grant McCready a work permit, so he couldn't legally hold a job or get paid. To make money, he had to go into business for himself. Early on, he bounced from odd job to odd job until an idea struck him.

As a student, McCready had spent his time in a Catalan-speaking region of Spain. There are also tracts of France, Italy, and an entire country, Andorra, where Catalan, a Romance language, is the main tongue. The eleven million or so Catalan speakers in the world have unique ways of expressing the time. It's never *one-fifteen* for 1:15. It's *one-quarter of two o'clock*. Catalan speakers don't say *one-twenty* for 1:20. They say it's *one-quarter and five minutes of two o'clock*. "And it just gets stranger as you go around the dial," McCready says.

With a friend, McCready developed a design for a watch dial that expressed time in the same way it was spoken in Catalan. He scraped together a few thousand dollars and found a factory in France that would make the watches. After five months, the little watch company was profitable. Not long after that, the watches became a fashion sensation in Spain. Celebrities wore them. McCready issued new limited-edition designs that would sell out in days.

McCready, who had been a musician his entire life, kept his hobby alive while putting his watch venture together. He played with a blues band that began to build a flock of fans in Barcelona, and eventually landed a contract with one of Spain's major labels. But as the band started putting their studio sessions together, it suddenly broke up, the members' personal issues all exploding at the same time. The band sang its songs in English, but McCready had been recording tunes on his own in Catalan. Once the band's fate became clear, he went to the record label and offered up his solo work. They liked it and signed him on.

Two of McCready's songs went to the top of the Catalan charts. It may have been a small market, but the success branded McCready as a legitimate music force, while the story of his watches had become something of a marketing legend in Spain. That combination led to his recruitment by the company that had taken over the venues built and used for the 1992 Summer Olympics in Barcelona. They wanted a marketing director who could find new ways to fill the seats. The stadiums had become white elephants that burned through cash and held too few events to justify their operations.

McCready successfully courted big touring acts and concerts to play at the venues. But even when a big show came to town, it was usually just for one day; the onetime income was nice, but the venues needed something steadier. So McCready started convincing bands to open their tours in Barcelona, which meant his venues would usually host a week or two of rehearsals plus a couple of opening shows. The plan worked, as Barcelona became the hippest place in Europe for bands to launch their tours. But just as McCready got the venue company profitable, Silicon Valley's tech boom wave reached Spain.

"I had all these friends who were getting rich with Internet companies—or thought they were getting rich," McCready says. "It seemed like the thing to do."

It indeed was the thing to do in 2000, and McCready went to work as the head of marketing for an online music startup called Deo. The Swedish company fashioned itself as the first open marketplace for music. Musicians and bands could upload their music to Deo, where they could sell it directly to consumers.

Like many companies born in that era, Deo had been infused with boundless optimism and a large pile of money. And just like hundreds of other startups, Deo had misjudged its appeal and market. Few people knew what a digital file was, and those who did were likely getting them illegally through sites like Napster.

A year later, Deo ran out of cash and folded. The experience proved valuable for McCready as he got to spend a year at the crossroads of music and technology.

During that twelve months, he met a small tech firm in Barcelona that had developed an algorithm for analyzing the underlying structure, patterns, and architecture of popular music. McCready spent time with the company's engineers and came to the conclusion that the technology actually worked. He proposed forming a new company built around the technology that would pitch to musicians and record companies. They called it Polyphonic.

The algorithms behind Polyphonic work a wondrous dissection on the music they're fed. The particular science behind the company's algorithms is called advanced spectral deconvolution. Different sequences of Fourier transforms and mathematical functions break songs up, isolating tunes' patterns of melody, beat, tempo, rhythm, pitch, chord progression, fullness of sound, sonic brilliance, and cadence. Polyphonic's software takes this data and builds three-dimensional models with it. By looking at, instead of listening to, the song's 3-D structure, the algorithm compares the song to hits of the past in as objective a way as is possible. Putting a just-analyzed song on the screen with No. 1 tracks of the past shows a kind of cloud structure filled in with dots representing songs. The hits tend to be grouped in clusters that reveal their similar underlying structures. Get close to the middle of one of those hit clusters and, while you're not guaranteed success, you're in very good shape.

When he was in Barcelona perfecting the algorithm, McCready ran as many to-be-released albums through his bot as possible. It was these test cases that would reveal if the algorithm had any real power. The algorithm rated most of the unreleased CDs as ho-hum. But one, the algorithm said, contained nine likely hits out of fourteen total songs. Those are Beatles numbers. McCready could hardly believe it. Nobody had heard of this artist, which made McCready worry that the bot was wildly wrong. But then the album, *Come Away with Me*, was released, selling more than twenty million copies and netting its artist, Norah Jones, eight Grammy Awards.[2] Jones had found the clusters.

"Some people describe a hit song as a brain itch," McCready says. "And you scratch that itch by listening to the song over and over again."

The clusters were the itchiest spots, and McCready thought his company had struck upon the formula for identifying musical gold. The music industry already tried to pick its own hits, but it was only right 20 percent of the time. McCready's tool, if it worked, would be the industry's holy grail.

THE A&R BOT SAID I DON'T HEAR A SINGLE

Polyphonic had company in its quest. A startup that called itself Savage Beast was out to do the same thing, although it employed starkly different methods. Instead of sorting through new music with an algorithm, Savage Beast employed hundreds of musicians to listen to songs and categorize them based on four hundred different attributes. The company first tried to sell its product to brick-and-mortar music stores such as Sam Goody, Tower Records, and Best Buy. But that plan tanked with the advent of rampant MP3 sharing. After the dot-com crash, Savage Beast ran on fumes for years, not even paying its employees.[3] Then in 2005 it changed strategy, using more algorithms and fewer humans to decide what people should hear, raised a little money, and got a new name: Pandora. By July 2011, the company was trading on the New York Stock Exchange, valued at $3 billion.

Pandora started with a different paradigm, and so too did McCready. His first hack allowed artists such as Ben Novak to upload his song for fifty dollars and get instant feedback on its potential via Polyphonic's algorithm. Record companies and talent scouts could then log in and browse the highest-scoring free agents on his site. McCready was trying to fix, or at least improve, a part of the music industry that had bedeviled both artists and record labels for decades.

Tom Petty's 1991 song "Into the Great Wide Open" features a well-known verse that refers to the record industry's equivalent of a baseball talent scout: the A&R man. A&R (artists and repertoire) staff at record labels are the gatekeepers to the recording industry. They can make careers and launch a musician from obscurity to stardom. In the pop music world, A&R people's jobs depend on finding singles. An artist without a single, as Tom Petty's sardonic invoking of that cliché suggests, isn't worth much.

One problem is that A&R is an inherently subjective trade. It's not baseball; a 100 mph fastball to one scout is a 100 mph fastball to another. There's no denying that kind of raw talent that can be quantitatively measured so easily. With music, however, talent exists everywhere. But not all musical talent has the potential to appeal to mass audiences. Some of the most brilliant artists in the world may never become known outside a small circle of fans. Yet other artists whose musical talent may not be deeper than that of an average eight-year-old piano player can take over the world with one catchy pop lick.

For that reason, music remains a business much like book publishing. The record labels depend on one album out of fifty to keep them profitable and justify the signing bonuses they dole out to new acquisitions. It's only by casting a wide net that labels assure themselves of scoring the hits they need. A&R people who unearth more diamonds than average are well compensated. Those who consistently pluck stars from obscurity become legends; these rainmakers often ascend to become leading executives at the label.

Even though A&R people are always hunting for promising artists, most of their signings come through personal relationships or direct referrals. The music business remains far from a meritocracy, even if merit is measured by the ability to hook teenage girls' ears. But what if A&R could be made into a science? Being right just 30 percent of the time would be a giant improvement on the industry's historical rate. And artists with a knack for pleasing listeners wouldn't have to wait for that random connection or recommendation that may never come. Polyphonic's algorithm, McCready thought, could prove the answer.

Novak's success primed Polyphonic for what seemed like a high ceiling. McCready's tool also identified Maroon 5 as an act carrying a high probability of success before the public had any idea who the band was. The software certainly wasn't right about everything. It's given high marks to heaps of songs that never gained traction with wide audiences. But there was no denying that McCready had created something that worked, something that could shape the future of the music industry.

But despite the promise of the technology, A&R personnel weren't too keen on giving credence to a tool that, if it lived up to its claims, would threaten their jobs. Many A&R people and recording executives laughed at the notion that a machine had any place in their world. When told about the work and ideas behind Polyphonic, Lorraine Barry, the global marketing manager at Virgin Records, scoffed. "The modern-day A&R man—a machine, a computer program? A bit of a frightening thought," Barry said. "I think it's a marketing ploy. It's pretending that it can be a science."[4]

Whether the software has made the A&R game into a science will remain debatable. But there's no denying that McCready's crew wasn't welcome within the industry. The music business isn't renowned for being open to change. "I think it finishes just ahead of the Amish in that respect," McCready says.

The business model behind Polyphonic depended on the music industry utilizing it as a new A&R instrument. That bet proved cheeky. Polyphonic, for all its wizardry, wasn't able to make any money. A&R people were loath to use a method that could hasten their own demise. Without their cooperation, Polyphonic floundered. McCready laid off staff and thought about what he'd do next after going from Nebraska farm boy to watch mogul to pop star and now tech founder.

With little to lose, McCready changed his model. In 2008 he moved to New York and became friendlier with the music industry. He recapitalized with new investors and dubbed his company Music X-Ray. Just as before, he invited artists to upload their work to his site and databases, but now he also allowed A&R men and producers to post veritable help-wanted signs when they might be looking for a new tune or artist.

Music labels, advertising firms, marketers, and music producers are often looking for a certain kind of sound. For instance, a music label may be on a mission to find the next Radiohead, or a marketing firm may think the Rolling Stones' "Brown Sugar" is the perfect song for their television spot, but they can't afford to pay the kind of money the Stones command. Unsurprisingly, when a legitimate talent scout at a major label issues a query looking for new artists, there's usually an avalanche of responses. The same goes for when a movie producer posts a request for an original score or a particular kind of song for a sound track. As people in the music business will tell you, they're very busy. Wading through thousands of submissions from random musicians—many of them mediocre or worse—isn't something that will often crack their daily agenda. This is why established artists tend to get the lion's share of new work. Finding new musicians takes too much time.

This is where McCready's algorithm comes in. It can quickly sort the right sounds from the wrong ones, allowing a music industry insider to find the closest match to their original query. In the instance of "Brown Sugar," the algorithm would comb its databases of submitted music for the tracks that best imitate the riffs, beat, rhythm, style, and overall sound the Rolling Stones struck in that song.

Or A&R people can simply look for the highest-scoring songs within different genres from the artists who have uploaded to Music X-Ray. McCready's warehouse of data grows larger each week. It's quickly becoming an encyclopedia of world musical talent, a heartening development for musicians out there, such as Ben Novak, who, as hard as they try and as talented as they may be, fear their work and sound may never make it out of their garage. It's more than possible that many of our future music stars will be produced by Music X-Ray's algorithm. It's already happening, in fact. Since 2010, McCready has landed more than five thousand artists opportunities with music labels and other commercial outlets.

One such musician, Tommy Botz, was a sixty-three-year-old recovering alcoholic who stumbled into Music X-Ray on the Web. Botz had been living in a rehab center for seven years, during which he wrote some of his two hundred songs, some of which he decided to upload to Music X-Ray. McCready's algorithm spotted Botz's music as an original take on Johnny Cash–like tunes and connected him with a Michigan label whose artists cut several of his songs.

Botz isn't an outlier. Lynne Ferguson, a grandmother of six, got the attention of the Atrium Music Group with her song "Tears to Gold" on Music X-Ray. The label picked up a total of eleven of Ferguson's songs and put out an album, Ferguson's smiling face right on the cover. Indie acts have found success through McCreary's bot as well. World Live Music & Distribution discovered Dominic Gomez on Music X-Ray and immediately signed him to a deal and a recording budget for hundreds of thousands of dollars. Movie studios are fishing the waters too. The Spike Lee–produced 2011 film *You're Nobody 'Til Somebody Kills You* plucked three songs for its score from artists found on Music X-Ray.

In exchange for putting the two parties together, Music X-Ray gets a small piece of the transaction. The revenue is starting to add up. McCready has rid his business of red ink and built something that the music business has embraced. At the end of 2011, there were fifteen hundred record labels and other professional acquirers of music using his algorithm, including those from almost every major house, including Columbia, Time Warner, Geffen, and EMI.

"I'm finally getting love letters from record labels," McCready says. There will always be human decision makers at some level, he thinks, but his bot and its feel for the clusters of popularity will eventually change who the public ultimately hears.

The efficiencies and the new breadth of artists that McCreary's model opens up to the music industry are such that it's only a matter of time until the major labels—all labels, really—come to rely on an algorithm to pick the musicians they sign and the songs they market. It's akin to when word processors first hit the market. At first, most people kept banging on their typewriters, as only the early adopters could see past the processors' small screens, funky printouts, and the scary idea of keeping all of one's work on a five-inch floppy disk rather than on paper that could be seen and held. But eventually, screens got bigger, the software got better, and the idea of using anything else became nonsensical. That day is coming to the music world.

Will that be a good thing for music? It certainly won't be a good thing for the average A&R man. But what about artists who are off the radar or not well connected? Will algorithms bring us a better Top 40? The answer is probably no. Algorithms may bring us new artists, but because they build their judgment on what was popular in the past, we will likely end up with some of the same kind of forgettable pop we already have. It's a clear foible of the technology that all these years of so-so music are included in its analysis.

McCreary, though, thinks otherwise. He hopes, as any entrepreneur would, that his algorithm will be at the center of the music world in the coming decades, but he says that its presence will lead us toward hit charts of more variety, not less. Artists with alternative sounds whose songs actually land near hit clusters because of their just-below-the-surface beats, melodies, and cadences will now be found rather than dismissed, he says. McCready can also help producers and artists move their work closer to hit clusters by tweaking rhythms, beats, and tempos. It's likely that one day we'll see garage bands jamming out a track and then scrambling over to a laptop screen to see how that version fared in the 3-D world of hit clusters. Such quick affirmation in a creative field is rare. But it also begs the question: rather than an explosion of variety, will algorithms lead to a music world of forced homogenization?

It's already true that a large chunk of the hits that populate the Top 40 were written by the same group of people. Martin Sandberg, for one, a Swedish songwriter who goes professionally by the name of Max Martin, got his start in the 1990s when he wrote a series of No. 1 hits for Bon Jovi, the Backstreet Boys, and Britney Spears. Since 2008, he's written more than ten No. 1 hits and more than twenty Top 10 singles, including "DJ Got Us Fallin' in Love" by Usher and "I Kissed a Girl" by Katy Perry.[5] Knowing that particular humans are gifted at writing hooks for the masses—and knowing what Mike McCready's algorithm already knows about the general characteristics of hit songs—it's easy to speculate that popular music could soon be ruled by bots. It's a certainty that record labels will serve up whatever the tastes of the day happen to be—and little could be better suited for such a task than an algorithm tuned to spin out saccharine hits.

THE BACH BOT

Hit clusters don't just apply to pop music. They're relevant for the masters of yore as well—Beethoven, Mozart, Bach, and Handel. McCready's software shows how some of the most popular symphonies fall into clusters of their own. While Mozart and Beethoven had no concept of the gadgetry or the algorithms McCready uses to figure this out, they most certainly understood that some patterns and rhythms had more power to appeal to the public. It's why some classical pieces sound similar. One success by one musician begets imitation by others, all of them chasing that same kind of sound—and acclaim. So if algorithms can determine what makes music popular, be it pop or classical, is it possible that they can take the next leap and simply create new music designed to please? There are those who have good reason to believe that algorithms will come to compose music as fine as any human. Chief among those believers is University of California at Santa Cruz music professor emeritus David Cope.

Cope uses algorithms to create symphonies, operas, and oratorios. His music—or rather, his algorithm's music—has become so good that it scares some musicians and has others wondering, where is the line between human and machine creation? At what point does an algorithm become an artist?

That's a question that Cope has been trying to answer since the 1980s. He's now getting ever closer, he thinks, to a conclusion that won't please those who would prefer a clear divide between what algorithms and humans create. This muddled situation isn't helped by the fact that Cope assigns his music-creating algorithms human monikers. He named his first bot Emmy.

When Emmy produced orchestral pieces so impressive that some music scholars failed to identify them as the work of a machine, Cope instantly created legions of enemies who would give his work little due and, in some cases, would fight against Emmy's pieces' being performed. At an academic conference in Germany, one of his peers walked up to him and whacked him on the nose. Shouting matches often erupt when Cope's work is discussed in the highest circles of classical musicians. Some of his colleagues at UC Santa Cruz call Cope "the Tin Man," after the heartless *Wizard of Oz* character.[6]

Cope's bot can string together notes that weave in and out with the power of Beethoven or the finesse of Mozart. That a machine can produce things of such beauty is threatening to many in the music community. "If you've spent a good portion of your life being in love with these dead composers and along comes some twerp who claims to have this piece of software that can move you in the same way, suddenly you're asking yourself, 'What's happened here?'" Cope says. "I'm messing with some very powerful relationships."

Emmy was the culmination of years of arduous learning, building, and rebuilding for Cope, whose hideout in Santa Cruz, a laid-back beach town, seems far away from the algorithmic revolutions spreading from the tech world. In reality, however, the only thing separating Santa Cruz from the densest collection of algorithmic smarts in the world is a strip of coastal mountain range, which Route 17, as it courses from Silicon Valley to the coast, disposes of in thirty miles.

While many in the music community scorned him, Cope found acceptance and accord in the computer science world. That musicians would brush away Cope as a heartless scientist isn't without irony. The California professor is a musician before all else. Cope's first memory is of lying on the floor, gazing up at the bottom of a grand piano as his mother worked the keys.[7]

When he was still in grade school, Cope became hooked on composers like Tchaikovsky, Rachmaninoff, and Stravinsky. At the age of twelve, he was enthralled with the prospect of writing music himself. He eventually headed to Los Angeles for graduate school in music composition at the University of Southern California.

After USC, Cope landed a teaching gig at Miami University in Oxford, Ohio, in 1973. The job allowed him time to do what he always felt destined for: composing. He began to find his style and the world liked it. Cope's original works were performed on some of the highest perches of classical music, from Carnegie Hall to the Kennedy Center for the Performing Arts. His scores even spread internationally, being played by orchestras from Peru to Poland.

Cope's love affair with musical bots began with research on a book he was writing about contemporary music. He had planned to include a chapter on the new field of computer-generated music, but he felt he knew too little about computer composition. So he chose the thorniest path he could, deciding to make music on the computer himself rather than simply interviewing others on the process.

Cope certainly wasn't the first to put algorithms to work making music. Guido of Arezzo in 1025 invented a method to automatically convert text into harmonies. He devised a set of rules that could take a seemingly meaningless jumble, like written words, and convert them to something bearable on the ears.[8] In the seventeenth century, Joseph Haydn, the man who wept at the brilliance of Handel, conjured a game called Musikalisches Würfelspiel, wherein rolled dice determined measures in a piece. A bevy of masters, including Mozart, dallied with this method.[9] In the 1950s, two researchers at the University of Illinois,

Lejaren Hiller and Leonard Isaacson, used the ILLIAC computer, at the time the fastest supercomputer in the world, to create what they called the *Illiac Suite.*

But to use modern machines to create music in the way he intended, Cope found no user manuals. So he busied himself with learning Fortran, Pascal, and Algo, the programming languages of the day. To compile his code and check his work, he fought for time on a room-sized IBM mainframe. The computer would spit out, in order, punch cards peppered with holes denoting the work Cope had done. He would then translate the numerical punches into pitches and notes before finally taking the score sheets to a piano to find out just what he'd created.

"I ended up producing an absolutely dreadful piece of choral music that I'd rather not talk about," he says.

Dreadful or not, the piece had been composed with the help of a computer processor, so Cope became a legitimate authority on the still young subject of algorithmic composition. He wrote his chapter and finished the book. Creating something new, innovative, and intriguing within the academic field of music is hard to do, and the book drew the attention of his peers across the world. He fielded dozens of calls about taking professorial seats elsewhere. The most persistent rings came from the music department at UC Santa Cruz, which got Cope out to California for an interview and convinced him to move in 1977.

The composer-programmer settled in at his new college and was soon working at full capacity, teaching and writing music that continued to harvest acclaim. In 1981, however, Cope's prolific ways were stymied. He had been contracted to write an opera, but found his brain's well of music creativity dry. He stared at piano keys and blank pages of paper for months. He would have just given the money for the opera back, but "I had already spent all of it," he says. He had four young boys and a modest paycheck.

At a loss, Cope thought he could perhaps end the stalemate by including a computer in his efforts. So he began sitting in on computer science courses at Stanford and UC Santa Cruz and spending long stretches in the library, poring over programming books that he would only begin to understand on the second pass. At some point in his studies, he became convinced he should be programming in a language called Lisp, one of the most advanced languages in computer science.

"Nobody told me it was tough, so I didn't know it was tough," Cope says.

He had begun the project with the goal of replicating his own composing style, so that he could complete the opera in a fashion his client would expect. But he found that mission amorphous. What was his style? It was a difficult thing to program. So he decided to build his program to mimic the more regimented style of the ultimate master: Johann Sebastian Bach. A year into his hacking project, his program was able to create chorales that, while they bested his effort from the 1970s, "weren't very good," he recalls.

The music professor kept at it. Cope built systems that sampled hundreds of Bach chorales and "learned" the patterns Bach used when composing his most moving pieces. The work was tedious. Each nibble of music his algorithm sampled needed to be formatted into smaller data pieces that made sense to the computer. The music-sampling technologies of today weren't even whispers in the early 1980s. To get the music into computer code, Cope would translate each note and chord by hand from sheet music into raw numbers. Each note required five numbers of code, representing time, pitch, duration, channel (instrument), and loudness or amplitude. Translating one piece could take weeks. Cope would often spend six months transcribing data with the goal of netting one set of results.

Breaking down Bach's brilliance into mere digits and pages of computer garble affected Cope. Could the genius be simplified to such a level, he wondered? Could there really be a Bach algorithm? As his program created more and more music, the answer could be interpreted as deflating or wonderfully affirmative, depending on one's perspective. The work of Cope's algorithm began to take on characteristics of Bach—the patterns, the melodies, the congruent measures.

But something wasn't right. The music made sense. The chord changes occurred when the ear expected. But the music lacked some essence of vitality. For all that it did right, the music didn't carry the vigor that normally imbued Bach. Cope transferred even more music into the database of the algorithm, hoping that the variety would help it break through. He studied his code and rewrote it time and again. But nothing seemed to help. The music remained listless.

Building a program capable of human-level composition seemed as distant as ever. Cope had been driven to the edge of a nervous breakdown on several occasions. Convinced an impasse was intractable, he would walk away from his algorithm for weeks at a time, swearing he was done. But he always found himself drawn back to the computer screen and into the thousands of lines of code he'd so laboriously assembled.

Out of the house on an errand one day, Cope was hit with a notion. Perhaps the problem wasn't his sample size or the set of rules he'd written—maybe the problem was that his algorithm followed the rules too well. Great composers certainly create by building on the patterns of the past, but they become exceptional by breaking the rules their music usually obeys. The Bach algorithm never broke rules, hence it never made music that had the capacity to surprise or hook the listener's ear. Cope dove back into Lisp, writing loops and random functions that would maintain patterns and structure in his music until unexpectedly breaking rhythm to lend the spark of an advanced human composer.

Completing this task represented the last strand of DNA for Cope's algorithmic composer. The resulting program knew how to create rhythms, measures, intros, and endings. It could mix slow, gentle notes with the brash rumblings of a baroque finale. Cope had devised a pile of code, a machine, that truly captured the spirit of someone as brilliant as Johann Sebastian Bach.

Sensing he was nearing some kind of ending seven years after he started, Cope named his creation Emmy, a play on the acronym for "Experiments in Musical Intelligence." On a day in 1987, Cope did what he often did—he put the algorithm to work just before he left the office for lunch. When he returned to his computer, Emmy had created five thousand chorales that closely traced the style of Bach.[10]

The Bach pieces were first performed at the University of Illinois's main campus in Urbana. As the chorales' performance drew to a close, the audience sat in a stunned hush. Surely this couldn't be the work of a machine! The crowd marveled. At a music festival in his hometown of Santa Cruz, Cope was approached by dozens of admirers who were blown away by the music. Unlike the Illinois crowd, they hadn't known Emmy wrote it. When they found out, there were yells, complaints, and, again, stunned silence.

Cope had still never finished the opera that originally sent him down this road. He fed Emmy most of his work from the 1970s and the algorithm began to shape an opera in the style of Cope. Teaming up with the algorithm, he wrote the opera in a week—after six years of being stuck. He turned it in and two years later it was planned, rehearsed, and performed in Richmond, Virginia. Nobody at that time knew of Emmy's role in the music. The opera garnered the most compelling reviews of Cope's career. "A supreme dramatic moment, punctuated by the captivating beat of drums," said the *Richmond Times-Dispatch*.[11]

This turn of events opened up a whole new set of questions: who truly created the music—Cope or Emmy? The professor has struggled with this question, but ultimately, he reasons, you don't credit the shovel with digging the hole, you credit the digger. Yet he admits that he sees both sides of the argument. "The question of authorship is a difficult one to answer—and it's one that's plagued me from the beginning," he says.

Encouraged by what was now earth-circling fame within the artificial intelligence world, Cope sought a recording contract for Emmy. What should have been a slam dunk proved immensely difficult. Classical music labels told Cope they didn't do modern music. Contemporary recording labels said they didn't record classical. When the record deal finally came, Cope could find no high-level artists who would play it, so he was forced to route it through a Disklavier, an automated piano that plays by electronic instruction.[12]

The album, *Classical Music Composed by Computer,* sold well and, just as important, landed Cope a lot of work. The recording included faux Bach as well as imitation Beethoven, Chopin, Rachmaninoff, Mozart, Stravinsky, and even a Scott Joplin piano tune.

The album also succeeded in getting Cope some of the respect musicians wouldn't grant him when he was seeking artists to record. Soon, famous classical performers were reaching out to him, hoping to play whatever Cope considered Emmy's preeminent piece. But such requests posed a problem: Cope had no idea which concerto to offer up. Emmy had now created several thousand pieces of similar quality. This alarmed musicians. Why would they want to perform a piece that was merely one of many? Artists wanted something of a guarantee that they would be playing Emmy's best. But Emmy could, with no warning, create something even better tomorrow. Musicians didn't want to play something so easily commoditized.

This is a subject that has threatened to ostracize Cope from the musical herd. He argues that all prolific composers, from Bach to Mozart to Beethoven, dealt in commodities. They developed patterns, Cope says, that they based on music they had heard throughout their lifetimes. Bach was never creating something wholly new, says Cope, but simply building incrementally on those who came before him. And he would repeat those steps with similar methods of patterning and building when he created anything. That's creativity, insists Cope—and it also exactly describes what Emmy did. This assertion that the classical masters were no better than pattern-spouting algorithms elicited the kind of response from the music world one would expect: rejection of Cope, Emmy, and his work.

But Cope would recapture the advantage. In 1997, with much of the classical music world paying attention, Emmy went head-to-head against two formidable opponents at the University of Oregon. Steve Larson, a professor of music theory at the school, helped put the event together. He also was one of the combatants. An audience of a few hundred people listened while three piano pieces, all meant to sound like Bach, were played, one composed by Emmy, one by Larson, and one by Bach himself. The audience then voted on the identity of each composition. Larson's pride took a ding when his piece was fingered as that belonging to the computer. When the crowd decided that Emmy's piece was the true product of the late musician, Larson winced.[13]

Douglas Hofstadter, a professor of cognitive science at Indiana University, oversaw the competition among Larson, Emmy, and Bach. Hofstadter's 1979 book *Gödel, Escher, Bach: An Eternal Golden Braid* contemplated artificial intelligence and music composition and was awarded the Pulitzer Prize.

"Emmy forces us to look at great works of art and wonder where they came from and how deep they really are," Hofstadter said at the time.[14]

With a victory in a controlled, academic setting, Emmy's profile soared, as did Cope's. But as the requests and praised poured in, so did the scorn. A machine couldn't bring the kind of soul to music that a person could, the critics kept on blaring. One detractor, Jonathan Berger, a composer and researcher at the Center for Computer Research in Music and Acoustics at Stanford, said Emmy knew where to put the notes, but it couldn't give music the true essence of a human composer. "There is a big jump between what Cope calls signatures and I call style," he said.[15]

Daniel Dennett, a philosopher and cognitive scientist at Tufts University, appreciated the marvel of Emmy, but still didn't think a machine could compete. "As wonderful as Cope's work is, there is something thin about it," he said.[16] A composer built from software has no sense of the real world and therefore no idea how to express true feelings through music, Dennett reasoned.

The dismissive attitude many held toward Emmy began to grind on Cope. He thought people heard what they wanted to hear. If people knew they were listening to music composed by a set of computer algorithms, Cope found they wouldn't allow the music to affect them. The listeners would say the music was well composed, but missing some intangible wisp of soul. But when Cope would play the same music for people and not inform them it had sprouted from the processor chip of a PC, the listeners would often revel in the sound and remark on how deeply the music touched them.

The power of a biased subconscious has been well documented; fighting against such natural tides exasperated Cope. Even worse, as far as he was concerned, people began to view Emmy as some kind of exhibition, like a concept car at an auto show that every attendee knows is just for show. But Cope's concept wasn't a far-fetched dream. It changed the way he viewed music and creativity. Algorithms like Emmy are the future of music, he says. Human composers will exist as they always have, he explains, but they will evolve to use every tool at their disposal—including algorithms.

Even as Cope attained superstar status within the artificial intelligence community—he gave keynotes at computer science conventions around the world—he lamented Emmy's lack of acceptance within musical ranks. As much as he'd made himself into a preeminent computer scientist through two decades of work, Cope remained a musician first.

The professor hadn't set out to create a novelty. He grew tired of people asking if Emmy could complete Schubert's unfinished Eighth Symphony or Mahler's Tenth—or if it could conjure what a Chopin opera would sound like (the composer himself never wrote one). "I began to feel I was viewed as simply a programmer, somebody who downloaded files and put together databases rather than someone who was involved in creativity," Cope says.

He also worried that Emmy's effective immortality affected people's perceptions. "Human composers die," he says. Their works become interesting and important because they become rare. Emmy's work seemed like it was never going to become rare because she could compose forever and create so many pieces that "we just drown in them."

In 2004 Cope followed his gut and destroyed Emmy. He erased all of the algorithm's databases, many of which he had arduously put together by hand. Emmy would never again create a series of imitation Bach keyboard inventions or Beethoven symphonies. Cope preserved the original code for teaching, because it had proven to be a true leap forward in algorithmic creativity. But by destroying the data crops that fed Emmy and enabled her to so well emulate human composition, he had ended the experiment and given his critics little to complain about. The world of algorithmic music, which had been creeping ever closer to that of humans, had been silenced.

It would prove a temporary quiet.

What Cope really wanted was a retreat from the spotlight and his critics. When he destroyed Emmy, he had already begun, in secret, on a second algorithm project. He dubbed his new creation Emily Howell. The algorithm is the direct, though more advanced, spawn of Emmy. Howell runs on the same kinds of logic, tricks, and equations. But she's better. Howell is the MacBook Pro to Emmy's Apple IIe. During one performance of her music in Santa Cruz, the audience listened rapt, unaware that the composition came from a computer. A professor whom Cope knows well said the score ranked among the most moving he'd heard in his life. Several months later, the same colleague attended one of Cope's lectures on Emily Howell and immediately discounted her music, saying it had no soul.

Soul or not, Emily Howell has composed eight operas as well as a handful of piano concertos. Howell's first record, *From Darkness, Light,* elicited reviews hailing it as brilliant, as well as those that fingered the music as atonal and shallow. Not wanting to tempt the same results that he did with Emmy—too many pieces, too little respect—Cope put an end to Emily Howell's short composing career following the work she did on her second album in 2011.

Not one to rest after three decades of pushing algorithmic limits, Cope has yet another, more dynamic cyborg composer in the works. As he develops his latest, most advanced music-making machine, he's chronicling it in a book entitled *The Transcendent Machine.*

Cope's new muse is named Annie. Instead of giving Annie a set of rules by which she must compose, as he did with Emmy and Howell, Cope has developed Annie to write her own rules utilizing what's called machine learning. There are different kinds of machine-learning algorithms, but Cope's dance on the edge of technology. Simpler machine-learning programs can be told desired outcomes and they then learn the best ways to get there. Scientists use them to tease out nonintuitive relationships inside vast stores of data. As they ingest more data, the algorithms get smarter. For example, Winestein is a machine-learning algorithm created by New Zealand's Smart Research team that recommends wine pairings for the dinner you're putting together. Winestein got its smarts by scanning thousands of menus with food-wine pairings from the best restaurants and sommeliers in the world. The more data it was able to gather, the more nuanced and smart Winestein's recommendations became.

What Annie does, however, is a step past Winestein. Annie learns how to learn. She decides what the biases will be, what the criteria are, and, ultimately, the path she takes to making harmonies. It can lead to some crazy music.

"The really interesting thing is that I have no idea what she's going to do sometimes," he explains. "She surprises me as much as anybody."

Cope has built Annie's algorithm to go beyond music. Anything that offers large volumes of past work that can be organized and sampled is fair game. In 2011, Cope published a book of two thousand haiku entitled *Comes the Fiery Night*. Annie wrote some of the book's haiku; the balance is composed of classics by Japanese masters. Haiku traditionally include three lines of verse with five, seven, and five syllables, respectively. The short-long-short rhythm pleases the ear in English, but it's even more powerful in Japanese, in which haiku must be written with the utmost efficiency because seventeen syllables can typically convey far more in English than in Japanese.

The book does not distinguish which poems belong to whom. Nor will Cope say what percentage of the book came from his algorithm.

"I've learned there's little advantage in telling people which are which," he says.

Here are three of the haiku, with at least one poem from a human and one from Annie (Cope wouldn't tell me who wrote what):

Cannot the fervor
Over the river's ending
Be postponed to now?

Water snakes writhe,
Water splashes in their wake,
A frog heads for home

The fire crackles loud
In the morning air as I
Wander on the path[17]

Which haiku are human writing and which are from a group of bits? Sampling centuries of haiku, devising rules, spotting patterns, and inventing ways to inject originality, Annie took to the short Japanese sets of prose the same way all of Cope's algorithms tackled classical music. "In the end, it's just layers and layers of binary math," he says.

Cope says Annie's penchant for tasteful originality could push her past most human composers, who simply build on the work of the past, which, in turn, was built on older works. Cope offers this without emotion, but he still harbors bruises from the music elite's reaction to Emmy. He won't admit it, but he seems ready for a fight.

The arguments will be honest on both sides. The specter that the greatest music ever created could be broken down into pattern recognition, rule making, and controlled rule breaking is just too much for some musical types to take. Artists tend to be people who define themselves by their work. They perceive their creativity as what separates them from the person in the cubicle, the salesman on the road, the MBA doting on a spreadsheet—the banal masses. Cope suggests that this artisan shield of superiority can be reduced to a long, efficient equation rather than some mystically endowed and carefully cultured gift. The man whose own creativity altered how we think of artificial intelligence inadvertently subverted thousands—if not millions—of others' stakes in the creativity game. If a composer's role to society can be filled by an algorithm, then what are humans left with?

But as bots move into the business of music creation, the door will be left open for disruptors. And whereas disruption usually comes from technology, it's likely that the pop charts, with what will certainly be a backbone of algorithmically conceived songs, will be left vulnerable to indie artists who create something truly different. Is it possible that an algorithm could have given us, based solely upon the past, Nirvana's second album, *Nevermind,* which spawned a new genre of music and bands? Or consider Outkast's "Hey Ya," written by Andre 3000. The 2003 song was like nothing most people had ever heard before; it's hard to imagine an algorithm drawing on popular songs of the past being able to conjure that. It's almost impossible. People take "Hey Ya" for granted now because they've heard it a thousand times, but that song was a true outlier in that despite being utterly different from anything else out there—a trait that usually disqualifies music from achieving widespread popularity—it was still able to become the anthem of American radio for half a year. Almost everybody, indie snobs and bubble-popping teens alike, adored the song. It's that kind of tune that requires true human invention. Cope may be on the road to such a thing, but he's still far away. And that's why there will always be a gap, albeit narrowing, for human artists to shoot for, even when algorithms rule the airwaves.

"Machines run on binary mathematics. Human's don't," Cope says. "Trying to replicate the human brain using binary mathematics is a lot different than creating it with cells and DNA. These are wonderful algorithms, but we may never do that."

Artists who compose jingles, pop hits, background music, and uninspiring movie scores should beware, however.

SOLVING BEATLES MYSTERIES

Just as algorithms are deciding what we hear in the future, they're also informing us of what we heard in the past. Some algorithms plumb the similarities of Bach and his contemporaries, looking for who inspired whom. Some algorithms look for clues that anonymous baroque-era pieces might belong to a great artist we already know. And then there's a set of algorithms belonging to a Canadian professor that has stirred up a bit of fuss within one of the largest and most ardent groups of music fans in the world: Beatlemaniacs.

Nearly everybody has an appreciation for the Beatles on some level. For those who don't swoon at the simple melody of "I Want to Hold Your Hand," there's a crate full of more potent songs. From the soul-scratching "While My Guitar Gently Weeps" to the poetic "Let It Be," the Beatles' innovation made them famous. Their breadth made them brilliant and their prolific catalog, completed in just seven years, shows an uncanny productivity.

Every aspect of the Beatles and their music has been studied, from the meaning of each verse they sang to the instruments they played to the women they dated. Their music evolved so quickly that acknowledging them as anything but genius seems irrational. Ebullient praise is common among the Beatle faithful. But Jason Brown is no normal Beatles fan. He's a math PhD who snaps off passable George Harrison licks on his guitar. More important, Brown has been cited around the world for his expertise in linking various sides of music to the granular math behind it. He also believes the Beatles—especially John and Paul—were math savants.

"If they had been inclined toward mathematics, they would have been among those geniuses that come along only every half century," says Brown. "They were the kind of people who solve all sorts of proofs that nobody else had. The odds that they would even be in the same part of the world and end up finding each other in the way they did—it's just amazing."

Drawing on his math and programming background, Brown began looking at some of the more prominent questions that have nipped at the Beatles' legacy for decades, beginning with one particular sound: the opening chord to "A Hard Day's Night."

The first chord of "A Hard Day's Night" may be the most recognizable rock strum ever recorded. Although it resonates for barely more than a second, even casual radio listeners know exactly what comes next. It's one of the more indelible sounds created by the Beatles. George Martin, the Beatles' longtime producer, acknowledged the effect and importance of the chord. "We knew it would open both the film and the sound track LP, so we wanted a particularly strong and effective beginning," he said. "The strident guitar chord was the perfect launch."[18]

But any good guitar player can tell you that reproducing that chord is impossible. Sheet music gets it wrong as well. Because nobody knew exactly how the sound of George's lead guitar chord was produced, different songbooks list different chords and notes for the song's opening shot. None of them are right. It took forty years for a mathematician with a computer and an algorithm to come along and solve the mystery of how the sound was made.

To be sure, there have been plenty of theories as to what created that particular chord. Many involved Harrison's twelve-string Rickenbacker guitar, which he'd been given on the Beatles' first U.S. tour in 1964.[19] But the twelve-string by itself wasn't enough. Many aficionados said they detected a piano in the sound. Others said they heard John's six-string and even Paul's bass.

Harrison himself half answered the chord question in a 2001 Yahoo! chat not long before he died:

a_t_m98 asks: Mr. Harrison..what is the opening chord you used for "A Hard Days Night"?

george_harrison_live: It is F with a G on top (on the 12-string)

george_harrison_live: but you'll have to ask Paul about the bass note to get the proper story.[20]

Getting access to Paul McCartney isn't within most fans' wherewithal, so the mystery remained intact.

Brown's own Beatlemaniac phase began when he was eleven, growing up near Toronto, when he got his hands on the Beatles' "Red" and "Blue" albums, a large compilation of the band's hit singles released just after the group disbanded. The music affected Brown and he immediately bugged his parents for a guitar. He practiced and played the instrument incessantly, often putting in eight- or ten-hour days during the summer of 1972. He played in bar bands all the way through college, where he considered majoring in music but ultimately picked math for its better career prospects. Years later, as a professor of mathematics at Dalhousie University in Halifax, Nova Scotia, Brown built an algorithm using what are called Fourier series to solve the Beatles chord riddle.

In 1807, Joseph Fourier published his heat equation, which described the distribution of heat over time within a metal plate that had been warmed from one point. Fourier's periodic function built on the work of Euler and Daniel Bernoulli to develop a series of trigonometric integrals—now called a Fourier series—that have proven incredibly handy in dozens of applications.

Among those who profited from Fourier's work were Fischer Black and Myron Scholes, who used a variation of Fourier's heat equation to create their algorithm that would go on to reorder Wall Street and earn Scholes a Nobel Prize.[21] An earlier Nobel was awarded to Herbert A. Hauptman and Jerome Karle in 1985 for their work utilizing Fourier series to model crystalline structures using X-rays.[22]

Fourier's work has also proved useful in decoding music, a fact Brown picked up in a math journal in the mid-1990s. Intrigued, he tucked the tip away in his subconscious.

In 2004, the Beatles were celebrated around the world for their fortieth anniversary. Brown found himself drawn in by the fanfare, which got him thinking about the opening riddle to "A Hard Day's Night." All the while, Brown's brain had been clinging to that little nugget from the journal article on Fourier series a decade before.

Instead of spending hours with his guitar on a quixotic mission to reproduce the chord, Brown thought, "Why don't I apply science to this problem?"

To understand what Brown was seeking to do, it's important to have a light grasp on the scientific principles behind music and sound. All sounds, like those we hear a band create, are made up of tones. Musical tones possess a frequency and an amplitude. The tone's frequency relates to its pitch and the amplitude relates to its loudness. Combining tones creates chords. To get the mathematical function of a chord, you add up the functions of its inherent tones.[23] The frequencies and amplitudes of tones can be modeled mathematically with sine and cosine functions, exactly like those used in Fourier series.[24]

When instruments are played, their recordings include both the pure tone of the instrument plus the smaller resultant sounds that all instruments emit, not to mention any other noises that may have been reflected into the recording microphone at that moment. By running the chord through an algorithm utilizing Fourier transforms, Brown thought he could sort the intended tones of each instrument contributing to the chord. Doing this would render an irrefutable answer to the forty-year-old riddle.

Brown started by transferring as pure a sample of the song as he could find to his computer. From there, he cut out a chunk of sound from the middle of the opening chord about one second in length. This little burst of harmonic sound would hold everything he needed to know. He ran the music through Fourier transforms he had composed for this purpose that disassembled the chord into its separate sound frequencies. The only problem was, the recording turned up 29,375 different frequencies. Those frequencies encompass the actual notes being played by the instruments plus background noise and side frequencies unwittingly emitted by the English musicians.

To focus on the correct tones, Brown sorted the thousands of frequencies by selecting only those with amplitudes greater than .02. Doing this thinned his list down to forty-eight frequencies; these would be the ones the Beatles were belting off their instruments. Brown then created a second algorithm to convert the frequencies to the notes played by the band.[25] His mathematical conversion assumed that the Beatles' instruments were in perfect tune. But the results showed that the Beatles had not tuned their instruments with any kind of rigor—not that it had any detrimental effect on their music or the song (as far as we mortal listeners would care).

With the forty-eight notes in hand, Brown began analyzing them. The first thing he deduced was that the three best-known transcriptions of the chord were incorrect. Each of these versions featured a low G2 (2 is the octave), which was absent from Brown's data. The Beatles played no such note. So before he had even solved the whole problem, Brown had debunked its three most popular answers.

One note, a D3, stuck out with an amplitude of nearly twice any other note. Brown was certain that Harrison had played his twelve-string guitar, whose sound can be heard definitively later in the song. But that instrument has consistently sized strings that produce notes of roughly the same amplitude, meaning it couldn't have produced the extra-loud D3. Nor would the note be from Lennon's guitar, for the same reasons. It had to belong to McCartney's Hofner bass.

But there were *four* D3s in Brown's data, so three of them (those with lesser amplitudes) remained unaccounted for. Even if one came from Harrison's twelve-string and another from Lennon's six-string, there still remained one more. On top of that was a pesky group of three F3 notes floating around. If Harrison had played one of them on his twelve-string, his guitar would also have produced an F4—but there was no trace of that note whatsoever.

Looking at his data, Brown was struck with an answer: the triple sets of notes erupted from a piano, not guitars. Anybody who has peered into a grand piano to case its innards knows that the instrument's keys bounce little hammers off strings to make the sounds we hear. But one key's hammer doesn't strike just one string. Three strings, in fact, create a piano's higher notes (with two strings for middle notes and one for the lower notes). This was a higher note, so a piano's strings would explain the notes in triplicate and the lack of an F4. Brown now knew that the chord included George Martin, who, as he often did in other Beatles' songs of that era, doubled on the piano while the band stuck to its core instruments.

Brown's algorithmic analysis turned up the real notes that Harrison had played on his Rickenbacker: A2, A3, D3, D4, G3, G4, C4, C4. This collection in no way resembles any of the sheet music floating about the world, which we now know was erroneously instructing people how to play this most classic of chords. Lennon hit a loud C5 on his six-string and Martin played D3, F3, D5, G5, E6 on the piano.

With that knowledge, Harrison's 2001 Web chat remark becomes all the more clearer: "It is F with a G on top (on the 12-string) . . . but you'll have to ask Paul about the bass note to get the proper story."

It was indeed an F with a G on top. And just as Harrison typed, the twelve-string played the G. Most people simply assumed, as Harrison probably wanted them to, that the guitar played the F as well. But the F actually came from Martin's Steinway piano, while Paul played a different note entirely. When Martin mixed the final track, Brown says, he made the amplitudes on the piano almost identical to those emitted by the twelve-string, hiding it behind the sharper sound of the guitar.

In 1979, looking back on his years with the Beatles, Martin alluded to the process that filmmakers undertake when they mix real footage with effects to create the illusion they want. "It shouldn't be expected that people are necessarily doing what they appear to be doing on records," he wrote.[26]

Professor Brown wasn't done with just dissecting the song's opening chord. George Harrison's guitar solo in the song, a brilliant blast of perfectly plucked staccato notes, had always seemed a smidge too perfect to Brown. How did Harrison play it so quickly? Brown had always thought it was possible that the solo had been recorded at half speed and then sped up for the final mix. There was little that pointed to his hypothesis being correct other than his gut.

Several Beatles experts had assured Brown that Martin and the band always noted on the recording sheets when changes in tape speed were used. There were no such notes on the song's sheets, however. If the solo had been recorded at half speed, others insisted, it would be well documented and well known. But Brown, emboldened by his breakthrough on the opening chord, kept on with his inquiry.

It was well known that the solo was actually a combination of Harrison on his twelve-string guitar mirrored by Martin on the piano. If the men had originally recorded the segment at half speed, they would have played it an octave lower than it sounds because doubling the speed of the tape results in the music kicking up an octave.

Because the piano was involved, Brown had a crack he could expose. So he sampled a tiny chunk of the solo and ran it through his algorithm. He chose a piece of the solo that included a G3 played by both Harrison's guitar and Martin's piano. When Brown examined the frequencies present in the recording, the G3 was indeed present in three frequencies. One of them would have been from Harrison's twelve-string. The other two were from the piano. This was the jackpot. There should have been *three* G3s from the piano had the solo been played at full speed.

As mentioned earlier, the higher notes inside a piano include three strings, while hammers for the middle notes strike only two strings. If playing the piece at full speed in the correct octave, Martin would have struck a G with three strings. Stepping down an octave on the piano, however, would produce a G note from two strings. So that was the answer—Brown had produced irrefutable proof that the tape had indeed been doubled up.

When he went back and looked at a BBC show from 1964 called *Top Gear,* which featured a live performance of the Beatles playing "A Hard Day's Night" about three months after they finished recording it in the studio, some inconsistencies Brown had noticed long before began to make more sense. Right before George's solo, the film hiccups from an obvious splice. The reason for this, says Brown, was because Harrison had not yet mastered playing the solo at normal speed, so the show dubbed that part in.

Brown is quick to point out that this studio work in no way detracts from Harrison's ability. Harrison did in fact go on to master the solo at full speed within weeks of the BBC recording. Brown wrote in a math journal, "It is to George Harrison's credit (and Beatles' fans' delight) that he had the nerve to play such a daring and quick solo at half speed, knowing that very soon he would have to step up to the plate and play it up to speed, with all the world watching."[27]

Brown has now set out to tackle another Beatles mystery, one that leads to delicate territory. The chord riddle has always been one that the Beatles and Martin enjoyed sustaining and, to some degree, explicitly acknowledging. Alternatively, many passionate Beatles fans know that both McCartney and Lennon claimed to have written the majority of "In My Life," a song recognized as one of the band's very best. Some, including *Mojo* magazine, say it's the greatest song ever recorded. It remains one of the very few tunes that Lennon and McCartney publicly disagreed on, a remarkable thing considering how much music the two created together.

To determine who was more likely to have authored the song, Brown built an algorithmic engine analyzing songs known to have been written by either McCartney or Lennon. From these song inputs, the algorithm builds a typical base structure of both writers' styles.

To do this, Brown utilized graph theory, which, as illustrated in Euler's bridge problem, employs edges and nodes. Brown instructed his algorithm to define repeating choruses as nodes and the verses in between as edges.

As of this writing, Brown says he's yet to feed "In My Life" to the algorithm. When and if he does, he'll compare its structure to that of the mean of both writers. From there, he'll make a declaration. Judging from the time that's elapsed on this project, however, Brown may be reticent to trespass on ground disputed by two of the Beatles, one of whom is still quite alive.

4

THE SECRET HIGHWAYS OF BOTS

THE VALUE OF ALGORITHMS IS all in their speed. If they weren't able to run through complex tasks in fractions of a second, they wouldn't be the revolutionizing force they've become. Speed is largely determined by one thing: hardware.

Neither Wall Street's warring algorithms nor the specter of a coming world that is increasingly ruled by algorithms would be possible without the galloping progress of hardware during the last several decades. Algorithms in our homes have been largely enabled by the wondrous code and telecommunications infrastructure that make up the Web. An algorithm that decides what movie you may want to watch, like the one employed by Netflix, is relevant because of the speed at which it can weight thousands of input factors correctly and get a decision back to you. If the process weren't almost instantaneous, it wouldn't be an effective tool.

Algorithms allow us to squeeze more and more into our days. Doing more in less time, in the algorithmic and computer world, is known as removing latency. At the cutting edge of this effort to eliminate latency is Wall Street, where the algorithm wars have become something of a contest to see who can do the same thing faster.

As Wall Street filled up with hacker talent throughout the 1990s, more and more firms employed similar strategies and similar algorithms. With so little separating their tactics and their code from others, traders continually pushed for better hardware, better computers, and better telecommunications lines. If traders couldn't outsmart their competition, then perhaps, the thinking went, they could outgun them. Even the best algorithms can be thwarted by speed. This is why, in 2010, two men built a secret gopher hole that stretched halfway across the country.

TO TRADE IS TO DIG

The unlikely tale begins in 2008, when a New York hedge fund asked Daniel Spivey to develop an algorithmic trading strategy that searched out tiny price discrepancies between index futures in Chicago and their underlying stocks and securities in New York. If the future cost more in Chicago than did all of its stocks in New York, the strategy would sell futures in Chicago and buy stocks in New York. When prices in Chicago and New York realigned—and they always realigned—Spivey's program would dump its positions and book its profit.

Spivey's strategy was no different than the arbitrage Thomas Peterffy did with similar stock indexes in disparate markets—except Spivey had to do it a lot faster. With the rise of algorithmic trading, any kind of low-risk arbitrage bet draws a crowd, which is why such strategies demand speed. If a trader or hedge fund can be first in to the trade, there's easy money to be made. For those just a millisecond behind, there's nothing.

To properly execute the algorithms he had composed for the hedge fund, Spivey needed space on the fastest fiber-optic network between Chicago and New York. His trade signals had to be among the first in line of the millions of trade messages firing back and forth between the markets. But neither Spivey nor the hedge fund could secure any bandwidth on the line; it was 100 percent full and defections were rare.

Spivey was able to find some space on what's known as lit fiber—shared space on a fiber cable that somebody else, such as a telecom company, owns and operates. It transmits information at the speed of light, but the information gets less bandwidth within the fiber strand and often faces more stopoffs and detours because a telecom company must satisfy customers in many places with many needs. But Spivey needed dark fiber, a strand of fiber-optic cable for which the buyer or leaser wholly owns the bandwidth. Dark fiber is called dark because it's essentially raw; whoever assumes control of the strand must supply the lasers, which can cost $5 million or more, to light the line and transmit the data from the beginning to the end. The information competes with nothing else within the strand; it's free to zip along at the speed of light wherever the strand may go. Dark fiber enables the fastest communication vehicle known to man.

With no dark fiber available to him, Spivey and his hedge fund client couldn't compete with the other investment houses and traders employing similar strategies. Their strategy and the code behind it, no matter how eloquently written, was useless. During the search for an available dark fiber line, Spivey had more than once remarked to himself and to others, "We ought to just string up a new line—it can't be that hard."

So he thought, "Why shouldn't I do this?" There were more pointed questions, however: where would the line go, how would he gain permission from landowners, who would lease space on the line, how much would such a project cost, and who would pay for it? Spivey had few answers. So he immersed himself into the arcane world of telecom, land rights, and high-speed circuitry.

Spivey, a native Mississippian, had started one of the first remote market-making operations on the Chicago Board Options Exchange. His programs traded options on the S&P 500 index, some of the most heavily traded securities in the world. Already an expert on algorithmic trading, Spivey spent months learning the finer points of transmitting data by light through a tiny tube. He soon was calculating in his head indexes of refraction and their effects on the speed of light. Light doesn't travel in an exact straight path through a fiber; it bounces from wall to wall as it courses its way down the line. The index of refraction expresses how quickly light bounces through a particular glass fiber. A natural math savant, Spivey quickly took to his new project, constantly scrawling on a notepad and uncorking comments on figures he'd just calculated.

Spivey traveled the territory of his imagined line. He talked to railroad companies who could lease him space in their rights-of-way for part of his pipe's journey; he talked to county boards that controlled highways; he talked to farmers; he talked to private landowners, golf courses, and environmentalists. The more he learned, the more he believed it was possible to build a new, private dark fiber pipe between Chicago and New York.

Most important, Spivey had devised a way to make his line shorter than existing paths of fiber between the two cities. These lines followed railroad rights-of-way for their entire route. Telecom companies like such paths because they require dealing with just one or two different entities, usually rail companies, to get between Chicago and New York. But the major rail corridors take a slight detour south in Pennsylvania and New Jersey. Spivey's proposed path would leave Chicago, nudge around the tip of Lake Michigan, and fire straight east toward New York with no detours.

The route would trim a hundred miles off the round-trip of existing shortest fiber paths. The speed of light in fiber, however, covers that distance in less than four milliseconds—four thousandths of a second. "That's close to an eternity in automated trading," says Ben Van Vliet, a professor at the Illinois Institute of Technology. "This is all about picking gold coins up off the floor—only the fastest person is going to get the coins."

Spivey knew his project could generate plenty of interest. If the line could be built, big algorithmic traders like Peterffy, Getco, Tradebot, and Infinium would pay large sums to ensure they were on it. The same went for banks like Goldman Sachs, Credit Suisse, and Morgan Stanley.

Using mapping software, Spivey highlighted county and state highways that hugged a straight line between Chicago and New York. Like existing fiber lines, he could use rail rights-of-way for the westernmost leg of the trip, as they stayed true to his course. That would save him some money. Laying new fiber in easy terrain like this runs about $250,000 per mile. But threading a pipe underground along the mountain highways of Pennsylvania and through the hills of New Jersey would cost three or four times as much. The project would run at least $200 million.

Spivey needed somebody with clout, money, and a tolerance for risk to partner with him on the venture. A fellow Mississippian came immediately to mind. James Barksdale had burnished his résumé as the COO of FedEx during the 1980s and as CEO of AT&T Wireless, which he propelled to the top of the telecom heap before he was tapped to be CEO by one of the most legendary companies Silicon Valley ever produced.

Netscape came out of the University of Illinois at Urbana-Champaign in the 1990s when students led by Marc Andreessen developed the first popular Web browser, then called Mosaic, which ushered in the age of the Internet. Modern browsers still carry much of the code Andreessen and his team produced while in school. And though Andreessen wrote the code and started the revolution, ending up on the cover of *Time* magazine at the age of twenty-four, it was Barksdale who helped Netscape become one of the hottest companies of the tech boom before AOL bought it in 1998 for $4.2 billion.

After the boom, Barksdale moved back to his native Jackson, Mississippi. He was essentially retired, but Spivey thought he might want to roll the dice on another zany idea. Spivey tapped his connections and got a meeting. Barksdale listened to the plan. He didn't know what to think.

"At first I said, 'Come on, you're pulling my leg,'" he recalls, wondering how the project would get through the mountains at a reasonable cost. But as he pondered it, he realized, "If it wasn't a tough problem, somebody else would have already done it."

Barksdale agreed to be the primary financer of the project, a big risk when no fiber contracts could be secured ahead of time because of the need for secrecy. Also disconcerting to Barksdale, he was unable to bounce the idea off many telecom people he knew because the project had to proceed in total stealth. If the construction became public too early, somebody, perhaps even a single trading house, could try to build their own line, quashing Spivey and Barksdale's exclusive speed advantage. Despite all the risks, Barksdale climbed aboard. His son, David, a mergers and acquisitions lawyer, became CEO and Spivey assumed the role of president. Spread Networks was born.

Spivey and the younger Barksdale immediately set out on a journey to secure permission to burrow through each and every land parcel their straight-line path required. "It's not always an easy pitch; some people don't want fiber optics in rural Ohio," says Spivey, who had to explain in exacting detail to county board members, some of whom still used AOL to dial into the Internet, why he wanted to dig through their territory. Some were suspicious of Wall Street, some were suspicious of the federal government, which Spivey had nothing to do with, and most were suspicious because they simply couldn't understand the point of a secret communications line.

By early 2009, the men had converted even the stubborn holdouts. They now had a clear path from Chicago to New York. By March, crews for Spread were moving dirt. The early going was excruciatingly slow. The contractors were used to controlled conditions, like those of open fields or rail rights-of-way. Spivey and Barksdale had them drilling, digging, and dredging through all manner of terrain and materials: mountain schist, valley shale, earthy mud, and Pennsylvania granite.

Exasperated with the pace, Spivey and Barksdale kept an ear to the ground for rivals who could usurp them. They feared that one of the large proprietary trading houses would find out about their dig and pull the trick off first. Even though the project's scope was huge, expensive, and almost unfathomable for a trading firm, players like Chicago's Getco had the cash and would gladly have paid to gain the advantage.

Spread's fears diminished as their crews became better in difficult terrain and the project's velocity increased without a competitor emerging. By the height of the project, Spread had 125 construction crews working at once. Some dug ditches in Indiana; others waded streams in Ohio or ran backhoes next to state roads in Pennsylvania. Workers in the Allegheny Mountains hung off the sides of cliffs, carefully placing blast charges that would remove enough rock to create ledges to hold their drilling rigs. Boring machines drove holes through prehistoric rock until their bits glowed white with heat.

Inside the insular world of trading, there hadn't been so much as a whisper of Spread's project. This was like George Washington crossing the Delaware in the middle of Christmas night. Spread would take the industry entirely by surprise. They only had to be first.

Spread's little hole ran for 825 miles, the shortest path possible between the 212 and 312 area codes. Information passing through this conduit travels at the speed of light, getting from New York to Chicago and back in thirteen thousandths of a second, shaving nearly four milliseconds off the previous fastest route.

Just before the final splice was made to the fiber in July 2010, Spread started pitching its weapon to trading houses, banks, brokerages—anybody with a lot of money and a thirst for speed. The reactions ranged from incredulous—*how did you do this!?*—to willing—*we're in!*—to those who felt less positive on the matter—*we will not be extorted!*

Any customer who spoke directly with Spread was bound to silence by a nondisclosure agreement. But nontrader observers registered their shock. "Anybody pinging both markets has to be on this line, or they're dead," said Jon A. Najarian, a cofounder at OptionMonster, which tracks algorithmic trading.

As soon as Spread made its final connections, customers queued up to get onto its lines. Anything that gives an advantage in common arbitrage strategies will always be valuable because arbitrage's low-risk profits will always be popular. Playing the referee in deciding who gets all of this riskless profit is speed.

As the path of lowest latency, Spread could command rates dwarfing those of other lines. Allied Fiber, which runs dark fiber through the longer railroad-plotted paths, asks $300,000 a year for its service. Multiple sources who negotiated with Spread said they were asking for eight to ten times the going rates, as much as $3 million per year. Spread won't divulge its total capacity, but its one-inch cable likely allows it to carry at least two hundred dark fiber clients. Even if some people aren't happy about Spread's high rates, the Spivey-Barksdale gamble is easily earning the cash flow it needs to justify its investment.

Spread Networks is merely the latest and most advanced weapon in an escalating war of hardware that began in the 1980s. As algorithms became the medium of choice for executing complicated strategies, their reach has expanded in tandem with the hardware on which they run.

While the advent of affordable PCs unleashed the potential of algorithms across the developed world, it was the Internet nosing into every facet of our lives that has made them all the more relevant. Determining the next field to be invaded by bots is the sum of two simple functions: the potential to disrupt plus the reward for disruption.

For a long time, that equation yielded the largest total on Wall Street, which is why so many of our smartest people, from engineers to physicists to PhDs, began flocking there. Still, that collection of brainpower didn't stop the industry from seeding economic disaster in 2008, nor has this corps of intelligence on Wall Street managed to solve any of the most vexing issues facing humanity. They just don't do that, even though by Goldman Sachs CEO Lloyd Blankfein's reckoning they're "doing God's work."

That Wall Street would bring the world to the edge of anarchy and then go whistling into the night is hardly surprising. That's a condition that may never change. But what does change, almost daily, is the hardware and technology available to grappling traders and their algorithms. The story of how Wall Street's technology has evolved is important because its progress eventually flowed to the rest of the economy. Even in the case of Spread Networks, a fiber-optic tunnel built for Wall Street, its effects have already leaked beyond the small world of algorithmic traders. Spivey and Barksdale's line now carries broadband to small towns that didn't have it. Spread is transferring large medical image files for hospitals and doctor's offices. The company offers these non–Wall Street entities lit fiber at affordable prices—an opportunity that exists only because algorithmic traders were willing to shell out millions for exclusive strands of Spread's dark fiber.

MONEY, SPEED, AND TECHNOLOGY HAVE ALWAYS GONE TOGETHER, FOR BETTER OR WORSE

The exclusivity of Spread Networks' dark fiber to those possessing barrels of cash may strike some as unfair and undemocratic, and perhaps it should. How can a free market be free when it takes $3 million a year just to lease the weapons required to play? But exclusive, expensive, and secretive technology has been part of the fabric of our financial and business networks for centuries. Those with the means to be fastest have an edge that enables them to make money. And those with the means are already rich. It's a foible of capitalism that is hardly new.

But without the pressure applied by our financial markets, there's no denying that our communications networks, be they word of mouth, pigeons, telegraphs, telephones, television, the Web, or dedicated strands of dark fiber, would have been developed at a slower pace. Without light-speed communication, there would be no cloud computing. Without cloud computing, which allows remote servers and computers to extend their processing power almost anywhere, all-knowing algorithms wouldn't be possible.

In 1815, Nathan Rothschild was one of the richest people in the world, and he regularly used his access to technology to add to his fortune. Along with his four brothers, he ran banks in London, Paris, Vienna, Naples, and Frankfurt. The brothers often used carrier pigeons carrying tiny rolls of encoded messages to communicate from bank to bank. This system afforded the Rothschilds' banks advantages over competitors who waited for communications via horse-bound courier.[1]

Pigeons have ferried messages for thousands of years across much of the world, from China and Greece to Persia and India. To intercept the messages of enemies, the Romans employed pigeon-seeking hawks. This led message senders to employ decoy pigeons as well as codes—all of which the Rothschild family used in the 1800s. In 1815, Europe was embroiled in one of its episodic melees, this one brought on by the return of Napoleon to power in France. Napoleon's French forces met the combined forces of Britain, Prussia, the Netherlands, and other allies in the battle of Waterloo on June 18.

On the bond trading floors of London, players waited for the arrival of news about the battle. A British defeat would send bond values diving for fear that the government would be on shaky ground. For Nathan Rothschild, this was the perfect situation to manipulate. His pigeon network delivered him news of Napoleon's defeat twenty-four hours before anyone else in London, including the British government.

Knowing victory belonged to the British, Rothschild headed to his normal spot in front of a pillar on the trading floor. When the session began, he immediately, with a flourish, started dumping bonds. Other traders assumed he knew something about the battle and that the news was bad. The market went into a free fall, just as Rothschild, who had stopped selling himself, knew it would. As the selling continued, dozens of Rothschild's minions began quietly buying up bonds. When the true news of British victory reached the masses the next day, the bonds Rothschild had purchased exploded in value.[2]

Technology always begins in the hands of a privileged few. Some of those who make the greatest fortunes, however, are the people who democratize technology. Such is the case of Israel Beer Josaphat, who grew up in Göttingen in the early nineteenth century and worked as a clerk at his uncle's bank, where he learned the fabric of finance. Josaphat emigrated to London, changing his name to Paul Julius Reuter. He came up with the idea of stationing pigeon fleets at the end of telegraph lines, which were starting to sprout across Europe. Reuter's pigeons would carry news of closing prices on the Paris Stock Exchange from Brussels to Aachen, one hundred miles away. His birds beat horse couriers and railroads.[3]

As telegraphs moved to more and more cities, Reuter began disseminating prices across distances of blocks rather than miles. His motto became "follow the cable," a mantra that still well applies to the global information network that Reuters has grown into. As services like Reuter's expanded, the information gap between the richest and the rest narrowed. Even in the case of Spread Networks, dozens of entities have climbed aboard the company's dark fiber, thus diluting the advantage that a sole proprietor could have claimed had the line been built by a large trading house or a bank. It's true that the high price of joining Spread Networks' service makes for an exclusive barrier to those who want to brawl with the biggest electronic traders. But Spread's availability still provides for a more democratized situation than existed in the Rothschilds' day, when their pigeon network was 100 percent proprietary and exclusive.

THE INCUBATOR OF SPEED

Following the Civil War, brokerage houses in New York were kept apprised of prices on the floor by runners who would sprint from one brokerage house to another, not unlike Peterffy's running men who delivered new computer boxes to the exchange throughout the day.

Beginning in 1867, the New York Stock Exchange began telegraphing market prices to ticker-tape machines. The first ticker machines were lethargic. "So slow even runners could view it with scorn," said the *New York Times*.[4] Nevertheless, the stock ticker system was the first electronic broadcast system on earth and would set the trend of technology debuting on Wall Street.

The Gold and Stock Telegraph Company, later absorbed by Western Union, owned the first network of stock ticker machines. The devices broke constantly. One particularly flummoxing break summoned to the trading floor a twenty-two-year-old mechanic who lived in the Gold Exchange's boiler room. After a few hours, the young technician was able to fix a problem that had stymied his senior coworkers. A year later, the same mechanic invented a newer, faster ticker machine that was installed across the entire U.S. network. The mechanic was Thomas Edison, and his system of ticker machines remained in operation for almost sixty years.[5] In the summer of 1929, just months before the October crash that preceded the Depression, the New York Stock Exchange and Western Union announced a $4 million program that would replace every ticker machine in the United States.

Anything less than a total overhaul of the system would be unfair, as the older ticker machines would leave those reading them at a disadvantage. In fact, as the new machines were installed in phases, Western Union had to devise a way to slow them down to run at the same speed as the older machines so they wouldn't offer unfair advantages to traders looking to front-run the rest of the market.[6] Not long after the new tickers were installed, Western Union announced that they would be raising the rates from $25 per month for the first time in a decade.[7] From there, the price of getting information as quickly as technology allowed escalated inexorably, culminating in our single strands of dark fiber that fetch $3 million per year.

Just weeks after Spread Networks completed its route, rumors popped up of a yet shorter pipe from New York to Chicago. The gossip, circulated on popular finance blogs and hacker chat rooms such as NuclearPhynance, posited that a company was drilling a hole in an absolute straight line through the earth from Chicago to New York. This hole would dispense with the earth's curvature, ensuring the shortest route possible.

As fun as it sounded, the project was a complete hoax. "Yeah, I heard about that one," says Spivey, the Spread Networks president, with a laugh. "Let's hope it doesn't get done anytime soon."

But with quants and hackers chasing them, Spread's castle will never be totally secure. In late 2011, a company called Cielo Networks started investigating constructing a wireless microwave network between Chicago and New York. Microwaves travel at the speed of light but aren't slowed by the refraction inside a fiber-optic cable. The method could theoretically shave four more milliseconds off Spread's 13.3-millisecond round-trip.

Luckily for Spread, there still exist large barriers to the scheme. Most frequencies that can carry enough data must be licensed from the government, and there's no knowing when such bandwidths will become available, if at all. And the money needed to buy such frequencies would be enormous. Not that traders have a shortage of cash, but they could find themselves bidding against the likes of Google or Microsoft, companies with pockets deep enough to ward off even Wall Street if they so choose. Even if the frequencies are secured, microwaves require clear lines of sight from transmitting station to transmitting station. Cielo Networks would in some cases need to obliterate mountaintops and secure towers that hew close to the true straight line between New York and Chicago, a task neither easy nor cheap.

Just as networks can be limited by geography, at least temporarily, algorithms can be limited by the hardware on which they run, at least temporarily. Hardware leaps in speed, size, and efficiency are as dependable as the sun's rise and fall. With these leaps, algorithms' utility expands in tandem. The phone in your pocket is more powerful than most supercomputers of the early 1980s; this little rectangle is the latest medium through which algorithms have infiltrated our lives, extending their power far past our desks and computer screens. The invasion didn't stop with room-sized trading machines on Wall Street, and it won't stop with your phone either.

5

GAMING THE SYSTEM

IN 1989, IBM CLEARED A few scientists to begin working on a machine whose stated goal was to beat the best chess players in the world. The machine, as many well know, was eventually dubbed Deep Blue. At the height of its powers in 1997, when it defeated the grand master Garry Kasparov, Deep Blue weighed 1.4 tons and ran 256 processors in tandem, which allowed it to examine 200 million different chess positions per second. Kasparov examined about three positions per second.[1] *Newsweek* declared the chess match "The Brain's Last Stand."

There was little elegant about Deep Blue's victory. This was a case of simply outgunning an opponent with more armaments. During a chess match, the computer explored nearly every possible scenario leading from a given move. It did this for any available move—no matter how useless or stupid the move might be. The bot then mapped possibilities for the opponent, and then again for itself, and then again for the opponent, and so on, resulting in a twisted matrix of moves with millions of possibilities. But Deep Blue was built to handle that load.

It took a large team of scientists years to build, simply, a bot to play chess. Was it possible to expect, coming off of Deep Blue's success, bots to delve into realms that required nuance and more of the unquantifiable elements that make up human intuition? In chess, all of the information is laid bare for both sides to see. There's nothing that's unknown, which made it a perfect game to be attacked with algorithms. And even that was hard. Beating the best humans at a game like poker, which incorporates bluffing, calling, folding—all moves that involve an unquantifiable and very human element—seemed impossible. But what's become clear in the years since Deep Blue's victory is that algorithms will continue to invade professions and skill areas that we have always assumed will remain inherently human. Chess was just the beginning.

In early 2011, IBM's newest creation, Watson, bested all human contestants on the game show *Jeopardy!*—including Ken Jennings, the most prolific champion in the show's history. That a bot could be so intellectually nimble in the way it processed random questions, speedily consulted raw stores of data, and issued answers was impressive. Whereas chess is a game played on a limited board with rigid rules, *Jeopardy!* is chaotic, arbitrary, and offers almost no guidelines on the content or nature of its queries, which can be pocked with humor, puns, and irony. To do it, IBM stored 200 million pages of content on four terabytes of disk drives that were read by twenty-eight hundred processor cores (the newish Apple computer I used to write this book has two) assisted by sixteen terabytes of memory (RAM). Watson ran so hot that it required two refrigeration units to cool its ten server racks that filled an entire room at IBM's campus in Yorktown Heights, New York. The whole contraption was built to run an algorithm that determined Watson's answers using more than six million logic rules.

But for all its abilities, Watson can't play poker with the big boys. Wall Street has always been fascinated with poker, a game that in many ways is a microcosm of human life. Poker encapsulates two of the most valued traits on Wall Street: a flare for bravado and quick analytical skills. Hush-hush games between trading titans have become legendary. A game featuring some of the most prominent hacker traders in the world, the Wall Street Poker Night Tournament, has become a premier event where people like Peter Muller, who built the algorithmic trading business of Morgan Stanley, lock horns with Ken Griffin, the billionaire hedge fund operator from Chicago. Muller won the 2006 game, and he'd had plenty of experience before then, including a stint on the World Poker Tour, where he won $100,000. Other Wall Street players include David Einhorn, the manager of Greenlight Capital, who placed eighteenth in the 2009 World Series of Poker and brought home $659,730, a staggering pile of money for almost anybody (except a successful hedge fund founder).[2] Even better, Steven Begleiter, the former head of corporate strategy at the now-defunct Bear Stearns, won $1.6 million in the same tournament. Wall Street's attention to poker was amplified by the game's general popularity spike that culminated in a purse at the 2011 World Series of Poker topping $64 million.

Just like markets, poker involves parries, fakes, smart traders, dumb traders, inside information, and, among thousands of other factors, the ever-threatening prospect of human irrationality. So it only makes sense that a computer science professor who tamed complex markets with some of the most advanced algorithms ever conceived would take on the world's most lucrative card game. Tuomas Sandholm, a computer science professor at Carnegie Mellon, has spent the better part of a decade working to break through on an algorithm that scares the best poker players in the world.

Creating an algorithm to effectively compete against the world's best card players is a complicated task. Making such an algorithm requires not only data crunching based upon the odds of winning a given hand with given cards, but also for the algorithm to acquire human qualities. It's not possible to succeed at big-time poker by simply playing the odds. That strategy can work in blackjack—as evinced by the teams of MIT students who took Las Vegas for millions—but poker requires a defter touch. To play at the highest level of poker, humans or algorithms must be effective at hiding big hands with small bets, and at other times declaring one's strength with large wagers to scare others out of the hand. The player, man or machine, must learn how to "limp in" to bets with weak hands for chances at unconventional, and thus sneaky, monster hands—like a full house of sixes and twos. None of these strategies can be used concurrently, of course. The best players pick when to bluff, limp in, or be demonstrative based upon the players they're up against. A world-class poker player must, above all else, learn how to use human psychology to his advantage.

At its highest levels, poker is one of the most nuanced and human games ever created. To make a poker bot worthy of the pros, Sandholm would need to construct an algorithm that made the leap from something doing millions of brute-force calculations per second (à la Deep Blue) to an elegant machine that could mimic human guile and effectively anticipate the rational—and sometimes irrational—strategies of its human opponents. If bots could tussle and beat the pros in this game, there exist few things beyond their reach.

TO BUILD CUNNING WITH CODE

Sandholm is an expert in game theory, the field of math concerning complex problems that include unknown variables and multiple parties with differing interests. Game theory has allowed Sandholm to build poker algorithms that make decisions with only partial solutions to their problem. "When you start trying to account for every single variable in poker, you get into trouble," he says.

In chess, every variable is known to all; in poker, the player only knows his own cards and the cards, if any, that are facing up on the table. Other than that, the player must rely on his own gut, his knowledge of his opponents' tendencies, and his ability to read other players' faces and hands for signs that they're bluffing or holding a good hand. Poker bots can't read faces, at least not yet, so to be effective against good players they must learn how to bluff, how to call others' bluffs, and when to fold in the face of what's likely a big hand for an opponent.

Sandholm, who speaks with the slightest of accents, was born and raised in Finland and looks the part of Nordic conqueror with his sandy hair, high cheekbones, and cleft chin. In 1990 he built his first commercial algorithm, an automated negotiation platform for ride sharing. Riders could introduce routes for which they sought rides, and drivers could answer with price quotes. Each rider introduced a different route variation to each driver, making for a complex market that could only be tamed by the perfect algorithm. Sandholm's foray into carpooling ultimately failed as the market didn't behave how his bot had expected. It turned out that some humans were perfectly happy to make a few bucks to drive others; others, however, demanded far higher—and thus, to the algorithm, irrational—premiums for the disturbance of their solo commute.

As he gained experience in complex transactions, Sandholm became better at modeling humans and their varying methods of decision making. He wrote game theory algorithms that negotiated markets with thousands or even millions of variables. For instance, when a company such as General Mills seeks to hedge its exposure to spikes in the price of grain, there are thousands of ways to do it, from buying call options to placing blind and bullish bets on commodities in New York and Chicago. Thousands of factors affect the decision, from weather patterns in North America and Asia to the direction of the stock market, the prevailing interest rates, and the threat of an insect pest or drought affecting the crop. Making the right bet at the right time—and for all the right reasons—can save General Mills millions of dollars. But many of the variables required to make the best decision are unknown. Some can be predicted quantitatively, although not definitively, such as the weather. Other variables, like how traders will react to certain events and how other large grain buyers will hedge their own positions, depend on how those players view their best interest. That self-interest is exactly what game theory helps predict.

With so many bettors and players in one market, the eventual direction of prices will be determined by the moves that all of the players make. Each player, in turn, does what she thinks is in her best interest. Game theory utilizes matrices with thousands and sometimes millions of nodes that account for each player's choices and their perceived payoffs. In a complex market, every decision made affects the decisions and the paths available to other players. Sandholm eventually built a company, Combinet, that constructed game theory software to help companies navigate all manner of complex markets. In 2010, Sandholm sold Combinet, by then a 130-person operation, to a private equity firm.

While solving problems with game theory for clients across the country, Sandholm was introduced to poker's mystery by other professors at Carnegie Mellon in 2004. An algorithm utilizing game theory, he thought, could prove effective in human strategy games like poker that contain unknown variables. In 2005, Sandholm and other CMU professors used game theory to write an algorithm that won a contest of Rhode Island hold 'em, a poker game with three cards in play rather than five. The bot was so good, in fact, that even pros couldn't beat it. But Rhode Island hold 'em, a small game for a small state, is a far simpler game than Texas hold 'em, the most popular game in high-stakes poker. In Rhode Island poker, there are approximately 3.1 billion possible situations that Sandholm's algorithms must consider.[3] In Texas hold 'em, there are as many possibilities as there are atoms in the universe, Sandholm likes to say.

In the Texas variation of the game, players draw two cards that they keep to themselves. Then another five cards are dealt up on the table. Players can make hands of five cards from any combination of their two and the additional five on the table. The nature of the game makes it especially suited to those who can read opponents. A player holding, say, a pair of kings might be riding high, betting large, until an ace is dealt faceup on the table. A cagey player at the same table with a next-to-worthless hand may bluff as if he's holding a second ace by placing a big bet. The holder of the kings, fooled by the big bet, thinks that he's been beaten by a pair of aces and folds. The best players alternate between playing strong hands loudly or softly, and occasionally bluffing with weak hands. To do this, players need to be good at two things: reading people and knowing the precise odds that they hold the winning hand. This is why building a world-class poker bot is so hard; algorithms usually aren't very good at predicting, analyzing, or gaming irrational human behavior. To a completely linear algorithm, bluffing a table full of players with a big bet while holding a seven and a two (the worst hand in poker) isn't rational. But there are times when such a play could make total sense to a pro. This unpredictable human element makes poker a field where only the most elite algorithm creators can play with success.

As Sandholm became fixated on the poker problem and began to dedicate years of his life to it beginning in 2004, he met others who were on the same quest. None of them were much of a match for his machines except those at the University of Alberta's Computer Poker Research Group, which had been hammering on the poker puzzle for twenty years. A friendly rivalry soon developed between Sanholm's Carnegie Mellon and the crew at the University of Alberta.

Sandholm's algorithm defeated Alberta's in 2011, giving it victory in two out of the last three years. Early in 2012, his bot outjousted every human it faced, even pros, when battling head-to-head in limit poker (where the bets have a ceiling). The bot slightly lags the best human players in the world when it comes to no-limit poker and poker with big numbers of people in a hand (more than four). In these larger games, bots are more prone to disruption by humans who bluff irrationally with bad hands to start but who may fall backward into a big card that saves them at the end of the hand. With more people in the game, there's a higher likelihood of somebody hitting a big hand on the fourth or fifth card dealt. In the case of poker hands that involve so many humans who, because they're human, can do something irrational at any moment, there's no algorithmic replacement for a lot of experience and a good gut feeling. Not yet. But considering Sandholm's progress thus far, the pros should be scared. Their Deep Blue moment is coming.

Sandholm's poker project has peeled back layers of understanding when it comes to predicting how and when scheming humans may make what seems to be an irrational decision with the goal of achieving an unlikely result—like bluffing a player who holds a pair of kings into folding. Predicting how humans perceive seemingly irrational actions to be in their own best interest is one of the last walls to be scaled by algorithms. If algorithms could sense irrational movements at the poker table, in the markets, or in simple everyday situations, money and lives could be saved. With that logic in mind, Sandholm's partner on the poker research project, Andrew Gilpin, has taken what they've learned from the behavior of humans in poker and applied it to the stock market with a hedge fund he started in 2010. His fund employs an algorithm that, just like in poker, is designed to keep others from smelling its true strategy.

GAMES OF THE PHYSICAL KIND

Long before poker became the game of choice among hedge fund titans and rising quants, there was sports betting. Trading pits exist to facilitate betting—and the betting always goes far beyond the financial instruments the pit is supposed to be trading. Sometimes more side bets are taking place in a trading pit than legitimate trades. When I was in college in the late 1990s—before it was easy to place sports wagers with offshore Web sites—one of my roommates would often gather big-game wagers from us and have the bets placed on the trading floor with a friend who worked at the Chicago Mercantile Exchange. That spirit has been transferred to a bot that now resides in a Las Vegas casino called the M Resort Spa, which specializes in a new brand of sports gambling modeled after the nonstop action of a trading floor.

The casino is owned by Cantor Fitzgerald, a Wall Street financial services company. Cantor believes its algorithm can make accurate odds on almost any sports event at any moment, down to the next three-pointer in a basketball game. Bettors in the M's sports book are given touchscreen tablet computers to carry with them as they roam the casino and watch hundreds of TVs tracking games. The M allows betting on virtually anything, like whether a batter will swing at the next pitch, or whether the next play in a football game will be an incomplete pass. The odds move every single second. Basketball bettors see different odds on their game eight times every minute.

Running all of this madness for Cantor Fitzgerald is an algorithm named Midas, which simultaneously juggles the odds on dozens of games featuring hundreds of subplots. Cantor built Midas with the lessons and expertise the firm gleaned from decades as a wily trading firm on Wall Street, where it was the first company to make its bond trading operations fully electronic. Midas runs from an expansive server room several miles north of the Vegas Strip.[4] To get the data necessary to guide its construction of Midas, Cantor bought Las Vegas Sports Consultants, an odds maker that had been handicapping games since 1982. LVSC's nearly thirty years of stats, results, and scores went straight into Midas's brain in a process similar to what Cantor and others use on Wall Street to create their trading algorithms.

Sports betting has long been a province dominated by gut hunches and casinos that set their lines according to how customers bet. If the money coming in on the Giants outweighs that of their opponent Patriots, Las Vegas casinos move their lines to make the Patriots bet more attractive. Equal bets on both sides minimize the house's risk and guarantee their 10 percent profit, what bettors call "the juice." Most even-money sports bets aren't even money at all; for every dollar wagered, the winner makes a profit of ninety cents—the remaining ten cents, the juice, belongs to the house. But Cantor wants more than the juice. The M isn't afraid to sometimes take naked positions on games—if the bot thinks it prudent. Cantor, with a team of fifteen hackers and its Midas bot, thinks it can become the dominant sports book in Vegas. Doing so will take years, but there's little doubt that the trading-floor-style betting at the M can be addicting. If the M finds huge success, expect Harrah's, MGM, Caesars, and the rest of the casinos on the Strip to start hiring the best hackers they can find.

ALGORITHMS AS CIA SPOOKS

While it might be wickedly cool to break out an iPhone or Android app that uses a game theory algorithm for strategy the next time you're playing the board game Clue—*if she thinks it was Colonel Mustard with the candlestick in the library, then I'll let the wretch keep thinking that*—Sandholm's breakthroughs in poker have opened doors for algorithms to attack countless other, more important problems beyond games—some that literally can decide life and death.

Not knowing the other guy's strategy—or the cards in his hand—makes decisions difficult in poker. The specter of the unknown also gives U.S. security forces problems when trying to hunt terrorists or when determining how to best allocate resources to disparate points of defense. It's also why fighting terrorism can be so expensive. But game theory algorithms are helping to change that. Sandholm's poker bot inspired University of Southern California computer science professor Milind Tambe to scheme against terrorism threats with algorithms. Taking cues from several of Sandholm's strategies inside the poker bot, Tambe developed an algorithm for Los Angeles International Airport that creates efficient and random security patrol patterns best suited to disrupt terrorism plots. The algorithm can examine what plans the terrorist might see as his best bet for inflicting damage through death, destruction, and chaos at the airport and then makes decisions to thwart attacks based on what it anticipates terrorists are thinking.

LAX started using the algorithm in 2007. The federal government was so impressed that it asked Dr. Tambe to build a bot that dictated patrols for Transportation Security Administration staff all over the country. A similar version of the bot went into testing at Pittsburgh International Airport in 2011. TSA has also begun to use the program to decide placement of air marshals on planes and the patrol times of bomb-sniffing canine units. The results have kindled work on bots that will one day make crucial battlefield decisions in places like Afghanistan, where convoy routes, temporary bases, and tactics must be designed to decrease the chances of ambush.

But before game theory algorithms took to defending airports and patrolling airliners, the CIA had come to depend on them like highly skilled intelligence analysts. When the Cold War thawed, there was a period of two years in the early 1990s, following communism's collapse, when the American and Russian intelligence teams were as chummy as they've been since the Second World War. The two countries hardly swapped secrets, but they did, for just a short period, quietly hit the pause button on what had been a fifty-year chess match. During this stretch the CIA and what used to be the KGB held sincere discussions on sharing information and intelligence systems. The Russians were keenly intrigued by one item the Americans showed off. It was a complicated game theory algorithm that had predicted the Soviet Union would crumble exactly as it had, sparked by an anti-Gorbachev coup that would ultimately fail.

The algorithm had been created by Bruce Bueno de Mesquita, a political science professor at New York University and a senior fellow at Stanford University's Hoover Institution. Bueno de Mesquita had been just another political science graduate student at the University of Michigan when he picked up a book by William H. Riker, *The Theory of Political Coalitions,* that used game theory as a template to predict politics. He became enthralled with the idea and brushed up on his calculus. He was soon improving on Riker's techniques, and by the mid-1980s he was the foremost creator of predictive event algorithms in the world.

When trying to predict how an intense political situation will play out, like the debate within Iran's leadership on whether to push through with the production of a nuclear weapon, there are thousands of factors that must be considered. The standard Fox News or CNN commentator would ingest all she could on the situation, consult her gut, examine similar events in the past, and then issue her prediction. Analysts at the CIA operate similarly, only with more information to work with than your average talking head. The fact remains that intelligence analysts at some point have to make a subjective call.

The algorithms that Bueno de Mesquita creates work differently—and they work quite a bit better. The CIA has hired him to perform more than seventeen hundred political and military predictions that span the globe and different geopolitical situations. In a study covering nearly twenty years of Bueno de Mesquita's work versus that of their own analysts, the CIA found that the professor with the game theory algorithms was right twice as often as the CIA's own experts. What does this say about geopolitical analysts? It's good that they have a future in television, because their role in the places that matter—the FBI, the CIA, the NSA—may be diminishing.

"When we depend on intuition, we miss a lot of the story," Bueno de Mesquita says. "Math gives us those answers."

Just like a poker bot, Bueno de Mesquita's algorithms consider the personal interests of every player related to a political outcome in a country like Iran. The only thing the algorithm assumes is that most people will act, pull, and speak up for the outcome that best serves their own interest. That being said, it also matters how important the issue may be to each player. While a supreme leader may prefer one outcome in a matter, the given situation may be of low importance to him overall, which allows for lower-level players, to whom the outcome may be more important, to exert more influence. So to run these predictive geopolitical algorithms, the user still needs intelligence data, but only certain kinds of data. They need to determine what groups or individuals are affecting the policy decision and what decision is best for each of the players. It's also necessary to know how much clout each player has and, just as important, how focused each individual is on this particular issue. Each party is looking to maximize their gains and minimize their losses—the move they believe will best allow them to do that is the move they will make.

That may sound like a mountain of intel, but it's less than what most CIA analysts consider. Intelligence services spend a lot of money gathering information on, simply, how we got here. They care about each player's history, their personal tales, their relationships, and their road to occupying the perch they currently inhabit. But the professor says this intel is immaterial—and his algorithm's success proves it. "If you walk into a room where a game of chess is going on and just look at the board, you can quickly understand what each player is likely to do," Bueno de Mesquita says. "How they got there doesn't matter—it's basically the same thing in the real world."

If we know each player's motivations, influence, and level of interest, we can set the algorithm loose. That's exactly Bueno de Mesquita's process. He spends days and days setting up the variables, the players, and the reasons why they care about this particular outcome, and then he lets the algorithm do the rest of the work.

As for Iran and the likelihood it will build a nuclear bomb, Bueno de Mesquita put together an algorithm on the matter in 2009. The bot weighed each factor: player, country, possible sanctions, and the abilities of the scientists in Iran. The bot concluded that by 2014 Iran will develop weapons-grade uranium—a fact that leaves many uncomfortable. But the algorithm also predicted that Iran *wouldn't* take that uranium and build a bomb. Instead, the country would keep its nuclear program a civilian one. The algorithm says that the important influencers in Iran will be sated by broadcasting their capability and wherewithal to create nuclear fuel. The leaders must weigh their desire to possess nuclear arms with the clear headaches—sanctions, Israeli military strikes, inability to sell oil—that come with them. Bueno de Mesquita says the algorithm probably recognized that many of the benefits of having nuclear arms—influence, respect, and regional authority—are also accorded almost as strongly by possessing a nuclear power program, something Iran is allowed to do under the international non-proliferation treaty. That conclusion, which at the time of this writing looks accurate, is far better than those put forward by most CIA and conventional analysts.

We shouldn't be surprised, though, that our experts are so easily bested by an algorithm. Berkeley professor Philip Tetlock conducted a survey of two thousand political experts and asked them to assign probabilities to certain events within their areas of focus. He then, in a control set of data, assigned random probabilities to each of the events. As the years went by, Tetlock collated the results, which, at least for the so-called experts, were disheartening. The predictions made by human experts were no better than those generated randomly.

Most CIA'ers are hired out of college, and because the old methodology stipulates that analysts should possess deep wells of knowledge of the region and people they're examining, most of the recruits have degrees in the humanities rather than science and engineering. For that reason, a lot of our intelligence focuses on personal histories and the drama that accompanies the entourages of people of power. These things might be interesting to talk about, but they're not essential to determining what humans in a far-off country will ultimately do. Our intelligence will home in on cultural incongruities, religious conflicts, and differences in majority opinions between races, countries, or people. But what really matters is what's best for those in control. Conveniently, this also makes for easier algorithm writing. If the seminal players and their interests can be well defined, and they usually can, even in the case of a country with leadership as stilted as Iran's, then an algorithm can give us better guesses on what people will do in the future than human analysts can.

The CIA is beginning to change how it builds its analysis teams, but it will be a long process. Jobs at the CIA are hard to get and the criteria by which candidates are judged—school, GPA, classic intelligence tests—produces a predictable result: an overabundance of East Coast Ivy Leaguers with humanities educations. The engineers and science majors who do apply are often at a disadvantage because their GPAs tend to be lower—a fact that's true across all of our university systems. It should be no surprise that people of certain backgrounds tend to favor those with similar résumés and reject theories that may displace their kind in what is one of the more powerful organizations on the planet. If game theory teaches us anything, it's that humans will almost always do what they perceive to be in their best interest.

But hard data drawn across twenty years can unseat even the stickiest dogma. There have been people inside the CIA and Department of Defense who have long pushed for bots to play more critical roles within the foreign policy machine in Washington. The most influential of the bot backers is Andrew Marshall, who was eighty-nine in 2012 and still occupied the job of director of the DoD's Office of Net Assessment. Marshall, who ascended to his position under President Nixon and has been reappointed by every president since, is a likely ally of algorithms, as he earned a graduate degree in economics from the University of Chicago, a place famous for its myopic focus on quantitative methods.

As for the Arab Spring, a sweeping event in world politics, Bueno de Mesquita had a line on that too. On May 5, 2010, he told a group that his algorithms had predicted that Egyptian president Hosni Mubarak would fall within a year, a result that came to pass on February 11, 2011. Bueno de Mesquita never made the prediction public because he gave it to a small audience, who after his speech put him on retainer and had him sign a nondisclosure agreement. But who would do that—and why? If you haven't already guessed, it was a large Wall Street investment house you've likely heard of. And why would they do it? To make the trade of the century.[5]

FOR ALGORITHMS, ALL IS FAIR IN LOVE AND BASEBALL

When Garry Kasparov lost the chess match with Deep Blue, it prompted George Plimpton, the late writer and actor of zippy wit, to dismiss the future of algorithmic takeovers as far-fetched. Plimpton himself had played chess with Kasparov, but he remained unmoved by Deep Blue's progress. "This is a very particularlized type of machine," Plimpton said. "Over the years, it may be that the machine is likely to be able to do other things. I'm not sure what, at the moment. It can't manage a baseball team. They can't tell you what to do with a bad marriage. They can't do any of these things."[6]

Before he died, Plimpton was able to witness the takeover of baseball by stat geeks firsthand. Some of the methods of these number hounds, who use algorithms to search minor-league rosters for hidden talent, have been well documented. The trend gained attention starting with the success of Billy Beane, whom Michael Lewis profiled in his 2003 book *Moneyball.* Beane made the Oakland Athletics into a perennial contender in the American League West Division by searching for players and lineups based on arcane groupings of complicated stats rather than the conventional ones that come on the backs of baseball cards. If the numbers worked, Beane did it, even if it led to some rather motley (read: chubby, short, and often bearded) collections of players on the field. The A's reached baseball's playoffs five out of seven years beginning in 2000 with one of the smallest payrolls in the big leagues. Compare that with the similarly cash-strapped teams in Kansas City and Pittsburgh, franchises that haven't been to postseasons since 1985 and 1992 respectively.

Oakland still sends talent evaluators to scrutinize prospects, but it also sets algorithms to sorting through the latest minor-league stats, searching for hidden gems. Beane's success could have been brushed off as luck, except that he kept getting the A's back to the playoffs after trading away some of the team's biggest stars for lack of money to pay them. Another heavy wielder of these new methods, Theo Epstein, led the Boston Red Sox to their first World Series title in nearly a hundred years in 2004, and then another in 2007. In 2011 the Chicago Cubs hired Epstein away in hopes that he could repeat the feat. The Cubs, despite their popularity, high ticket prices, and large payrolls, haven't won a World Series since 1908, the longest such drought in history.

Plimpton would be technically right in that a bot has not yet managed a major-league team from the dugout, but some human managers are already very expensive bot proxies. A growing cadre of baseball skippers have come to rely on computers and heavily sifted statistics to guide their decisions. For many clubs, it's not a manager's gut that decides which pitcher heads to the mound. It's an algorithm that tracks how well the pitcher has fared against hitters similar to those appearing that inning, recent workloads, the pitcher's ability to hold runners on base so they don't steal, their effectiveness at home versus on the road, night games versus day games, grass versus fake turf, and so on.

Baseball may be the sport most easily hacked by algorithms and the mining of arcane statistics, but it's hardly the only league susceptible. Such methods have become popular in the NBA as well. There's a growing divide between younger team general managers who make decisions with help from algorithms—like Rich Cho, the general manager of the Charlotte Bobcats, who has an engineering degree and five years of experience as an engineer at Boeing—and the old guard who prefer to roll with their gut. Sorting arcane stats with software is most helpful at evaluating midlevel players and identifying bargains. Nobody needs help to see that Kevin Durant and Blake Griffin are elite players. But even stars can be missed by human observers.

Take the case of Jeremy Lin, the second-year point guard whose NBA future was so tenuous in February 2012 that he was still sleeping on the couch of his brother, a dental student at New York University.[7] Lin, who played college ball at Harvard, went undrafted in 2010 and bounced around the league, getting cut from rosters as often as he played, before landing a spot on the Knicks early in the 2011–2012 season. At first he rode the bench. But then a string of injuries to New York regulars in February 2012 pushed Lin into the starting lineup. In six consecutive wins he scored 25, 28, 23, 38, 20, and 27 points, the fourth effort coming against Kobe Bryant and the Los Angeles Lakers. Bryant was impressed. "Go back and take a look, and the skill level was probably there from the beginning; it's just that we didn't notice it," he said.[8]

But an algorithm belonging to basketball analyst Ed Weiland did notice it—two years before any of this happened. It keyed on Lin's two-point field goal percentage in college, a lofty .598, and an arcane stat that Weiland calculates called RSB40, which reflects a player's rebounds, steals, and blocks per 40 minutes of play. Weiland considers this stat a good indication of a prospect's overall athleticism and ability to survive at the NBA level. The algorithm revealed that Lin's potential, on paper, was similar to past college point guards such as Allen Iverson, Jason Kidd, Rajon Rondo, Andre Miller, and Penny Hardaway—an illustrious bunch. In 2010, backed by his statistics, Weiland wrote, "If he can get the passing thing down and handle the point, Jeremy Lin is a good enough player to start in the NBA and possibly star."[9]

But Weiland is a FedEx driver with a stat blog; he doesn't manage an NBA team or even have the ear of people who do. Because of Lin's slim frame and the fact that he played in the Ivy League, not known for being a hotbed of NBA talent, he went undrafted and unnoticed. He didn't even play in earnest until February 2012. Even after a dip in production, Lin still averaged more than 14 points and 6 assists per game, remarkable numbers for an undrafted player with little pro seasoning. Without the use of stat-scanning algorithms, there was no way to know he was sitting out there, waiting to be discovered.

Data-crunching stat guys only started taking over in baseball within the last fifteen years at the highest levels of the game. But sorting out potential love interests by algorithm can be traced back to 1965, when Lewis Altfest, a young accountant, together with his friend Robert

Ross, a hacker at IBM, created what they called Project TACT (Technical Automated Compatibility Testing).[10] It was the first commercial dating service that ran on pure data, though a service just for Harvard students, Operation Match, anticipated TACT by one year.

Available to New Yorkers, TACT charged single city dwellers five dollars and asked them to answer one hundred multiple-choice questions that covered topics from politics to hairstyles. Altfest and Ross then took the surveys, translated them onto punch cards, and fed them into an IBM 1400 Series computer, which would produce five matches for each person. Within a year, five thousand people had signed up to find love with this first of dating algorithms. The service only lasted a few years before Ross, beckoned by the changing world of Wall Street and its creeping automation, left for a career in finance. He said he thought computer dating was just a fad.[11]

The fad did go dormant for thirty years, but in 2011 it was a $4 billion industry growing bigger by the day. One in six new marriages sprouts from connections made on Internet dating sites.[12] One of my former coworkers met his wife on eHarmony thanks to an algorithm with hundreds of variables and a proprietary base of code that's surely worth millions of dollars, if not billions. The dating Web site eHarmony matches people based on 258 questions that judge them on twenty-nine core traits. Users aren't allowed to browse for dates on their own; the algorithm decides all.

eHarmony's algorithm is based on research by Galen Buckwalter, who was a psychology professor at the University of Southern California before heading to the Web company. The site says it now has a hand in more than 2 percent of all marriages in the United States, which translates to 120 marriages per day.[13] Other algorithms are plying the waters too. Perfectmatch.com puts people together with an algorithm created by University of Washington sociologist Pepper Schwartz. Helen Fisher, an anthropologist at Rutgers, created an algorithm for Chemistry.com based on the neural chemistry of people in love, according to the company. OkCupid, yet another dating service, was founded by four math majors from Harvard who built a site that continually nudges users for more information that can be fed to OkCupid's algorithms. Users now compose most of the quizzes, which can be short or long surveys asking any number of questions, some of which get dumped into OkCupid's algorithm. As of 2012 the site had more than forty thousand quizzes.

But there are scientists who still don't buy that algorithms can bring us love any better than random bottle spinning. Research done by Eli Finkel, a professor of social psychology at Northwestern University, and Benjamin Karney, the same at UCLA, says that when the algorithms behind big dating sites cull the field for users down to a top set of matches, they do not outperform the compatibility potential of picking people out of the field at random. There may be something to that. OkCupid has analyzed thirty-five thousand couples from its site and discovered that, of its hundreds of thousands of questions, the one that best signals the potential for relationship longevity is, "Do you like horror movies?"

"Eighty years of relationship science has reliably shown you can't predict whether a relationship succeeds based on information about people who are unaware of each other," Finkel says.[14]

The reason for algorithms' failures, Finkel and Karney say, is that most relationships are subverted by issues that emerge only after the couple meets and begins spending time together.[15] How people communicate under stress and how they solve problems with others, the professors say, are far more indicative of how a relationship will fare in the long run. But there is little doubt that the creators of dating algorithms will search out ways to discern these exact traits the academics talk about. There are reasons to believe they'll eventually be successful. As Terry McGuire will show us in chapter 7, algorithms can already read our words for the true meaning behind them. They know what we're thinking.

6

PAGING DR. BOT

THE *PHYSICIANS' DESK REFERENCE* IS a book that combines pharmaceutical companies' prescribing information and directions for every drug they make. You've likely seen it in the background at your own primary care office. The *PDR* is updated annually; the 2012 version has 3,250 pages. Theoretically, doctors are supposed to keep apprised of what's inside the book, which includes thousands of updates every year on dosages, side effects, complications, and possible harmful drug combinations. Even the most pedantic studiers would have a hard time ingesting and retaining the information contained in its pages. To be fair, the book is a reference work to be checked when a doctor doesn't fully know or understand everything about a drug. But an algorithm acting as a doctor could take the information found in the *PDR* and commit it to memory in seconds, constantly updating itself with nuanced adjustments based on new data pushed out by drug manufacturers. While medicine seems to be rather unhackable on its face, the truth is that it's already been shaken by algorithms—and the real changes are yet to come.

THE ARBITER OF LIFE

Sitting in an audience in 2004, Tuomas Sandholm, the creator of poker algorithms, found himself riveted by a speaker and his topic. Al Roth, a Harvard economist, was talking to the World Congress of the Game Theory Society in Marseille, France. Roth was explaining the challenges facing kidney transplant networks in dealing with the circumstances facing each donor or recipient: blood type, location, health, age, relatives, and so on.

With tens of thousands of people sitting on waiting lists in America and across the world, maximizing the number of matches between donors and patients is the biggest mission facing the transplant system. The methodology used to get organs to as many recipients as possible has changed countless times during the last few decades, but thousands of potential matches are still unmet. The reasons for that are complex and diverse—but not so complex that they can't be fixed.

As Sandholm listened, he shifted in his seat when Roth began discussing how organ donor networks were experimenting with game theory to increase the numbers of transplants they made in a year. Roth himself pioneered the use of game theory to match up the more than twenty-five thousand annual U.S. medical school grads with hospitals for residencies. Before his system was implemented in 1995, the U.S. resident program was infamous for making nearly everybody unhappy, from hospitals that didn't get their top candidates to residents who were torn away from their spouses or shifted to hospitals that were weak in their chosen specialty.

Roth also created an algorithm to match graduating New York City eighth graders, who can attend more than seven hundred different public high schools, with the school that's the best fit according to the student's preferences and the school's rigor, acceptance policies, and location. Before his algorithm, the matching process was a mess. One Department of Education official compared it to a chaotic bazaar in the

Middle East. The program was so ineffective that only 66 percent of students opted in; now the Roth algorithm decides the futures of 93 percent of the graduating eighth graders in New York City.[1]

Following his hack of high schools, Roth began work to apply game theory to the problems vexing kidney donor networks—a far more complicated task. Changes within an organ transplant chain can happen fast, from the condition of a patient and the availability of an organ to the willingness of some participants to follow through. In Marseille, Roth explained how the quick pace of the problems affected his team's work. As the pool of transplant candidates grows larger, he told the audience, the complexity of the problem increases exponentially.

As Roth continued, Sandholm looked around the room, which was mostly full of economists. He started thinking that he was as good a candidate to solve this problem as anybody. "This wasn't an economics problem, I realized," he says. "This was a computer science problem."

When disease strikes at somebody's kidneys, causing them to fail, the person is left with two options: either find a person willing to donate one of their kidneys, which allows patients to live a relatively normal life in many cases, or, if a donor kidney can't be found, submit to dialysis, where a machine filters the blood of excess salts and toxins several times a week at four to five hours per session. The third option is death. More than 20 million Americans live with chronic kidney disease, and 350,000 undergo regular dialysis treatments.[2] Despite the fact that the United States spends $40 billion a year on dialysis treatments, more than 20 percent of U.S. dialysis patients die every year.[3]

The fact that eighty-six thousand people sit on the kidney waiting list makes abetting transplants a weighty mission. Of the twenty-five thousand or so annual transplants in the United States, more than 60 percent of them begin with newly deceased donors. The majority of the rest come from live donors related to the recipients.

In his 2004 speech, Roth spotlighted the promise of what's known as pairs matching, which was only responsible for several dozen transplants a year at that time. Pairs matching happens when a patient has a relative or friend willing to donate but, for whatever reason, usually incompatible blood type, the patient's body won't accept the donor's kidney. If a second incompatible pair can be found where the second donor matches the first patient and the first donor matches the second patient, then transplants can be made. Finding an opposite pair that satisfies all necessary matching requirements, however, is something of a long shot. Introducing more pairs to the mix creates more available matches, at least theoretically. But throwing thousands of pairs into a bucket and attempting to sort them out efficiently is complicated.

Before Al Roth started tinkering with new methods for matching pairs, they were put together by hand with little more than a spreadsheet. If a match wasn't patently obvious, it wasn't made. Before Roth's work, in fact, there were no national or regional pools for kidney matching anywhere in the United States. Roth developed a system that automatically found the most efficient way to put pairs from the pool together, but the algorithm was limited in the numbers it could handle, a problem he lamented at the lectern in Marseille.

Sandholm tracked Roth down after his speech and offered his help on the matching project, using some of the tactics he'd picked up in creating complex market algorithms and the world's most advanced poker bots. "The code needed for something like the kidney matching pool is very difficult to write," Sandholm explains. "It had taken me twenty-two years of experience with these algorithms to get to this point."

One of the problems with crunching big pools of pairs was the sheer number of possibilities and the computer memory such a calculation needed. Each candidate and donor carry with them more than twenty parameters that need to be optimally matched, including blood type, body mass index, viral history, willingness to donate or accept a left or right kidney, age, blood pressure, relationship to candidate, types of antibodies in the blood, and even where they're willing to travel and the difficulty of their match in general (acutely difficult matches are given extra precedence as they should be taken care of before more easily matched candidates). All of these factors make a computer run through millions of possibilities to find the best solution—and each calculation is more complicated than creating solution trees for a chessboard.

Before Sandholm's work, pairs matching couldn't be scaled up to a pool of thousands. But he wrote a new algorithm and coded it in a way that allows the computer to process the giant calculations without exhausting its memory. He tapped methods he used in making his poker bot, instructing his algorithm only to remember the most important bits rather than all of them. The algorithm can approximate the blanks when necessary. Increasing the pool size, and building a bot that can handle it, will save more lives, which is the whole point.

A national pairs matching pool, run by the nonprofit United Network for Organ Sharing, began using Sandholm's algorithm and taking pairs late in 2010. Once a month, Sandholm's algorithm sorts through the pool, looking to make the maximum number of matches. UNOS started publicizing the pool in late 2011 with the goal of getting its numbers to what Sandholm calls "critical mass."

As more people register for the UNOS pairs matching pool, which is also gathering participants from regional kidney transplant networks, the program will become increasingly powerful, leading it toward that all-important critical mass, where both hard-to-match and easy-to-match candidates stand a great chance of extending their lives. When the pool opened in late 2010, there were only 140 pairs participating, which led to only about a dozen matches from the first pass of the algorithm.

In a small pool, one tough match—due to blood type, location, lack of immunity, or a bevy of other factors—can lock up the whole system. That hard-to-match patient may come with a paired donor whose organ would fill a number of other difficult matches, but can't be used until the patient is also found a match. Such situations can prove intractable in small pools of just a few hundred pairs. But as the pool grows to include more than a thousand participants—Sandholm designed his bot to handle up to ten thousand—there's almost always a puzzle piece that fits, allowing others to fit in behind it. That's what Sandholm calls critical mass.

"Tuomas's algorithm gives these tough matches real hope of finding that exact right kidney—and that possibility gives them a reason to live," says Ruthanne Hanto, who runs the UNOS pairs matching program.

YOUR DOCTOR BOT

Wouldn't you like to have a doctor who will:

- Always be convenient and available.
- Know all of your strengths and weaknesses.
- Know every single risk factor your past conditions might signal.
- Know your complete medical history.
- Know the medical history of the last three generations of your family.
- Never make a careless mistake or write an incorrect prescription.
- Always be up to date on every new treatment and medical discovery.
- Never fall into bad habits or ruts.
- Know by heart each one of your baseline measurements: pulse, cholesterol, blood pressure, weight, lung capacity, bone density, and past injuries.
- Monitor you at all times.
- Always be searching for the hint of a problem, be it a heart tick, a creeping blood pressure increase, a cholesterol surge, or even trace changes in the air you expel, which could indicate early-stage cancer.

There exists no human doctor who can do these things. An algorithm can and will do all of them. In a small but growing number of hospitals, consulting an algorithm is task number one when diagnosing any admitted patients. The evidence in favor of algorithms at the bedside is piling up; the hospitals that use algorithms in their standard processes have fewer complications, more correct initial diagnoses, lower fatality rates, and, yes, lower costs.

Most of the public would likely be willing to accept an algorithm that metes out organs when it's clear that the bot saves many more lives and facilitates far more transplants than models that existed before. People will be less sanguine, however, when change affects their own doctor and regular medical treatments. In health care, the complaints, even when they're fueled by misinformation, fear, greed, or worse, can be shrill. Imagine the fervor that will come when some insurers suggest replacing physicals by real live doctors with constant monitoring by an algorithm. Algorithms are bound to replace doctors on some level—a prospect that isn't as scary as it sounds. When it comes to matters of health, though, many people's attitudes can be summed up thusly: change is bad; no change is good.

Still, people's resistance to computer-based algorithmic diagnoses and treatment is predictable. The United States already spends $2.6 trillion a year on health care, 18 percent of its GDP, a far higher percentage than our peer countries, where in every case people not only live longer but also suffer far lower infant mortality rates and generally enjoy better years at the end of their lives.[4]

Health care is full of easy wins for algorithms. The first thing to change will be how we analyze test results. The frequency, complexity, and costs of tests, from Pap smears to X-rays to MRIs to CT scans, have been among the major contributors to the explosion of health care costs during the last twenty years. When somebody shows up at the doctor not feeling well, they're often subjected to a gauntlet of tests no matter what their symptoms indicate about their condition. When that flurry of tests is over, another one often follows. Even when there's a

99.9 percent likelihood the test isn't needed, it's usually still mandated. The result, for those doing the testing, is an avalanche of easy revenue—thus exemplifying how capitalism, although it's an effective paradigm for most of the economy, is not an efficient way to administer health care. A multitude of stats can be called on to support this, but simply consider that the United States' average life expectancy of seventy-eight years is the same as that of Cuba, a country that spends less than 10 percent per capita of what the United States does on health care but achieves roughly the same result.

But this isn't about an ideological battle. It is about algorithms bettering the system we have in the United States—and they will do that initially by becoming the default scanners, observers, and analysts of all of these tests. Where once your results required the expensive attention of a radiologist or a pathologist, in the future the attention of an algorithm will do just fine. It will, in fact, do better.

Take the Pap test, originally called the Pap smear (named after its Greek inventor, Georgios Papanikolaou). The test has cut down on cervical cancer mortality rates in the United States by more than 90 percent since it was first introduced in the 1940s.[5] It examines a slide of a small sample of cells drawn from a woman's cervix. The slide first goes to a cytotechnologist, a person specially trained to do one thing: look for signs of abnormalities in the cells that may indicate cancer. Suspect slides are then passed to a pathologist, an MD who may earn more than $300,000 per year. There exist algorithms that can already replace the cytotechnologist. Even better (or worse, if you're a cytotechnologist), the algorithms help find more instances of cancer than their human competitors by scanning each image for visual clues that mimic those known to reveal cancers in old cataloged images. In one study carried out by BD, a medical technology company that creates some of the equipment behind the testing, labs using their algorithm in tandem with a cytotechnologist spotted 86 percent of cancer instances versus 79 percent for those scanning without the algorithm.[6] The rising use of algorithms in medical scanning will snare more problems, like cancer, earlier, and minimize false negatives, the scariest issue with such scans. Even incremental improvement would change thousands of lives; in the United States alone, more than fifty-five million Pap tests are carried out every year.[7]

In another case, algorithms bettered lone radiologists in their search for what are called lung nodules, which can indicate lung cancer. Doctors using the algorithm to analyze their MDCT scans, an advanced form of X-ray, improved an average of 16 percent in spotting nodules, the monitoring of which, if caught early, can spare patients from the middle and late stages of lung cancer, for which the survival rates are less than 15 percent. In another study, from the Elizabeth Wende Breast Clinic in Rochester, New York, using algorithms as diagnostic interpreters was shown to decrease the incidence of false negatives in mammogram scans for breast cancer by 39 percent, meaning that cancers that are normally missed by humans are spotted at a far higher rate by algorithms.[8] Stanford University's Cancer Institute already has an algorithm scan every mammogram following a radiologist after Stanford studies indicated that it led to more early detections of breast cancer.[9]

The first step in slowing out-of-control testing costs in U.S. health care was to ship some test analyses to well-trained doctors in India. The next—and final—answer will be turning all of this work over to algorithms that, when perfected, will do a better job than humans in India, America, or anywhere else.

The jobs for people reading scans, tests, and patient metadata are certainly endangered. Just as threatened as this workforce are the people who measure and dole out our medications. Pharmacists in urban areas can make more than $130,000 a year. Their job is stressful, detail-oriented, and can be, unfortunately for pharmacists, easily quantified. That means the profession begs for a bot invasion, and it's imminent. In 2011, the University of California at San Francisco opened a pharmacy staffed by nobody except a single robot. A Swiss logistics firm, Swisslog, created the $15 million robot for UCSF.

The machine, which receives information straight from the electronic messages that already go between doctor's offices and pharmacies, has long, dexterous arms that pluck and pack pills from thousands of bulk boxes that are built into the walls of the contraption. The bot receives all information regarding the patient, including their condition, any allergies, and all other medications they may be on. Algorithms within the bot quickly check for conflicts and complications with the new prescription, ensuring that there will be no detrimental drug interactions. The algorithms in the UCSF machine don't need any time to read up on the latest in pharmacology; they get updated drug information through electronic messages from pharmaceutical companies and assimilate it immediately. The machine, unlike a human, forgets nothing. That's not to say that a machine can't make an error—bugs can affect the best-written programs. But bugs are also easily fixed and algorithms can run concurrent tests on the bot that doles out the drugs, building safety redundancy into the robot pharmacist.

The bot in San Francisco has now filled two million prescriptions without making a single mistake. And there's no human contact between the pills and their packaging, eliminating the chance of contamination.

So how does the human competition fare? There's a strange dearth of data on pharmacy error rates, and the few extant studies show results that span from the scary—a 4 percent error rate—to even scarier—10 percent error rates.[10] It seems, however, that an accepted conservative number within the industry is about 1 percent. A national study of fifty pharmacies by the American Pharmacists Association showed that the average error rate is 1.7 percent.[11] All of these numbers are scary. There are 3.7 billion prescriptions filled in the United States every year, which means, even by the most conservative of measurements, that there are more than 37 million prescription errors annually. The American Pharmacists Association estimated the number to be even higher, at 51.5 million.

The costs of these errors to patients, our health system, and all parties involved in the pharmaceutical process are dear. Pharmacists are not helped by the fact that there remains a shortage of them nationally, which leads to overworking, understaffing, and even more errors. The demand for an algorithm-powered robot in this role is undeniable. Picking out pills, cross-checking for drug interactions, and ensuring that quantities and medications are exactly right are skill's made for algorithm-powered bots. Few tasks are so quantifiable. There's a business side of this equation as well. Walgreens, the country's largest pharmacy chain, had four jury-decided lawsuits involving fatal prescription errors brought against it during thirteen months in 2006 and 2007, with awards topping $61 million.[12] What if Walgreens could eliminate such liabilities from its balance sheet? It can and it will.

BUT WHAT ABOUT THAT DOCTOR'S APPOINTMENT?

Vinod Khosla, a cofounder of Sun Microsystems and as prescient a venture capital mind as exists, is one of the many bright people who believe that algorithms will make health care better. "Eventually, we won't need the average doctor," Khosla writes. Algorithms, he says, will provide much better and cheaper care for 90 to 99 percent of our medical needs.[13]

There are also compelling people such as Jerome Groopman, the Harvard doctor and splendid *New Yorker* writer, who disagree. Groopman says that relying on algorithms and evidence-based medicine is dangerous, that it alters how doctors think, making them less effective at diagnosing that odd malady that's already fooled twelve other MDs.[14]

But few doctor visits unspool like an episode of *House*. There's no reason why our most flexible diagnostic minds couldn't be consulted on seemingly intractable cases, such as the one of Anne Dodge, who as Groopman tells us had her life saved by an insightful gastroenterologist after dozens of doctors whiffed on her real ailment during the course of a hellish fifteen years. Dodge, who had been diagnosed with anorexia and bulimia, had been told to eat grains and carbohydrates in order to pack on weight after she slipped to eighty-two pounds. This recommendation was killing her. Gastroenterologist Dr. Myron Falchuk finally correctly diagnosed her with celiac disease, wherein the intestines can't properly process gluten, one of the main proteins in wheat.[15]

Why should we be taxing minds like that of Falchuk with routine procedures, queries, and cases that specialists in his realm face every day? We should save the tricky cases for solvers like Falchuk and give the rest—like the cold you had last week or that sinus infection from last year—to the algorithms. They're cheap, accurate, and they get smarter every day.

Many doctors are familiar with algorithmic medicine to some extent. Residents and even older MDs often consult diagnostic cheat sheets when assessing a new patient. But their utility comes at a price; another doctor/writer, Fred Herbert, a professor of medicine at the University of Texas Health Science Center, complains that the algorithms, when printed out on paper, grow labyrinthine and can take up a whole page. "They have so many steps with so many arrows pointing in so many directions that I ultimately give up trying to figure everything out," he says. That problem is easily fixed by sticking all of that information into an iPad app that only asks the doctor for inputs rather than showing him the guts of the entire operation on a printed page. But even that prospect doesn't impress many doctors. Herbert maintains that mathematicians are for managing numbers and doctors are for managing patients. Managing patients like numbers, he says, is a "deplorable trend."[16]

But there's accumulating proof that says otherwise. Intermountain Medical Center in Utah has used data and algorithms to improve that hospital's performance and make it one of the preeminent healing centers in the world. Researchers and doctors come from as far as Switzerland to view the methods used at Intermountain. It should be no surprise that the man leading Intermountain's charge, Dr. Brent James, spent his undergraduate years studying physics and computer science at the University of Utah before heading to medical school.

James has built a system of relying on data rather than mere doctor's intuition when making decisions about how to treat patients. Everything the hospital does is charted, analyzed, and reported so it can later be examined by an algorithm, from ventilator settings to stent use after heart attacks. Using data produced only at Intermountain, James has been able to rewrite procedures the hospital follows derived from the predicament of the patient. Algorithms search the collected data of previous cases, informing doctors at Intermountain on future diagnoses and courses of treatment.[17]

The results are hard to argue with: using its own data, Intermountain has cut the death rate for coronary bypass surgery to 1.5 percent, compared with 3 percent nationally; it has reduced the death rate of pneumonia victims by 40 percent, compared with national numbers; and its percentage of infants delivered preterm is far below that of the rest of that country. The trend of more preterm deliveries in the United States is well documented; according to the World Health Organization, it has increased 36 percent during the last twenty-five years. But the rate at Intermountain has actually decreased thanks to the hospital's treatment practices that have been informed by tens of thousands of cases it has categorized, quantified, and crunched. All told, James and Intermountain have increased successful outcomes for more than fifty clinical conditions, comprising more than half of the hospital's patients.[18] James's work begs the question: do you want a doctor—or would you perhaps be better off with a bot?

"Don't argue philosophy," James said of technology doubters to the *New York Times*. "Show me your mortality rates, and then I'll believe you."[19]

And that's just treatment standards at the hospital. Preventive care often starts in your physician's office, but algorithms can improve efficacy rates there too. Eventually, algorithms will start improving your general health right from your mobile phone. In 2010, Sequoia Capital, the most revered venture capital firm in the world, invested in a company called AirStrip, which after another funding round in February

2012 has raised more than $30 million in its effort to provide real-time patient data straight to doctors' iPhones, iPads, and other devices. Next up is home monitoring, wherein an app will monitor you at home and at work, constantly scanning your vitals and looking for early signs of trouble. Early versions of AirStrip's home app will be focused on improving outcomes for congestive heart failure, which affects six million people in the United States, according to the National Heart, Lung, and Blood Institute.[20]

The Dutch technology company Philips now makes a ninety-nine-cent app for your iPhone or iPad that takes several vital measurements of the person looking at the screen, using the device's camera. Philips's algorithms, by examining minuscule color changes in the image pixels of our faces, can figure heart rates to fine degrees of accuracy; to determine a user's breathing rate, the algorithm watches his chest for movement. Philips says it has been working on the advanced algorithms for the innovation for years, but it was the new hardware in the form of the iPad 2 and the iPhone 4S, introduced in 2011, that allowed the company to put these algorithms to work, for anyone, for less than a dollar. Upcoming apps from Philips and other tech companies will allow for instant measurement of blood pressure, temperature, blood oxygenation levels, and signs of concussion.

Coupling real-time monitoring from devices we already carry around in our pockets with detailed knowledge of our DNA, our genes, and what diseases we're naturally susceptible to will be one of the great medical steps forward of our time. DNA analysis isn't usually a customary step of primary care because it's time-consuming and expensive, just like most things in medicine. But this will change, as algorithms designed to map your DNA faster are put on the case. Armed with your genes and their variants, other algorithms will take preventive care to a new level.

Consider Renae Bates, a business development manager at an architecture firm in Australia. Bates was in the hospital for a tonsillectomy in 2006. The operation went smoothly and her doctors and nurses began to clean up, leaving her to rest on the operating table. But she couldn't rest—she couldn't breathe either. Bates is among the 5 percent of people in the world affected with pseudocholinesterase deficiency, which makes her more vulnerable to prolonged paralysis from anesthetic drugs used in operations. As she struggled for life on the operating table, one of the nurses finally noticed and the team restored her breathing with a tube. But had the hospital known Bates's condition ahead of time, the incident, which could have easily killed her or resulted in serious brain damage, could have been avoided.[21] Thanks to services like 23andMe, many of us will be able to head off such occurrences very soon. For $200, the company takes a saliva sample from you by mail and returns a detailed analysis of your DNA, its algorithm teasing out a variety of fascinating factors, from your ancestry to your health risks and potential reactions to medications. To be sure, some doctors and health experts say that 23andMe's tests offer no useful information and that consumers should save their money. And some states, including New York, have ordered 23andMe and similar services to get approval from the state's health department, declaring their tests to be medical and therefore open to regulation. Such regulation is "appallingly paternalistic," says 23andMe, adding that people have a right to information contained within their own genes.

Such genomic scanning is now fast and affordable, thanks in part to Nick Patterson, a Wall Street hacker who after eight years at Renaissance Technologies, the quantitative hedge fund, joined up with the Broad Institute, a joint research center of Harvard and MIT, in 2001. Working at Renaissance, which makes money off of sorting data and spotting patterns that nobody else can, made Patterson the perfect person to help the Broad Institute, which was drowning in DNA data so deep that the researchers there found it to be unnavigable. The information from sequencing just hundreds of people's complete DNA genomes produces data so copious that researchers usually don't send it to others across the Internet because such a transfer would take weeks. Instead, they FedEx boxes of high-capacity hard drives back and forth.[22]

Patterson composed algorithms to find, search, and sort patterns and relationships from the DNA data, changing the speed at which we can analyze DNA. His algorithms also uncovered one of the larger discoveries in evolution during this century. For decades, it was popularly theorized that humans and chimpanzees parted ways at some point: we went one way, they went another. But in 2006, Patterson's algorithms figured out that we actually got back together with chimps millions of years later, interbred, and then split again for good. The algorithms also led to the discovery that humans crossbred with Neanderthals less than one hundred thousand years ago. The tryst gave us what Peter Parham, an immunogeneticist at Stanford, calls "hybrid vigor," endowing us with a powerful immune system that allowed humans to colonize the world.[23]

A generation from now, algorithms like Patterson's will scan our DNA and tell us what diseases we're likely to get and even when they may come. Treating those maladies will be handled by a computer the world knows well: IBM's Watson.

When you head to the doctor's office with a health quandary, your appointment usually goes something like this: Your doctor asks a question, you answer; your doctor asks another question, you answer. This pattern goes on until your caregiver can suss out what she thinks is your exact problem. She bases her diagnosis on your answers, which lead her through a tree of possibilities within her head. An algorithm can do the same thing. It's also true that most algorithms don't converse; they don't weigh factors and nuances that may have affected your state, your answers, and your sickness. But Watson and the coming class of algorithms that will invade our medical world are different. They're imbued with the latest speech recognition capabilities, a dynamic sense of human feelings, and, with the advancement of face-reading algorithms, the ability to read what the creases in your brow are truly saying.

Watson will knit these kinds of capabilities with as rich an evidence-based diagnosis as any doctor could muster. A doctor may witness your initial symptoms when you walk into the room and, as any human would, assume the easiest explanation, thus coloring her examination, her questions, and her prognosis. Watson is unaffected by any of this. Your first several answers to the bot's questions open up thousands of possibilities within Watson's brain. It digs further with the questions that help it best eliminate or narrow the possibilities until, finally, it can issue an answer or put you on track for the exact tests that will reveal the answer. No willy-nilly tests, no gut feelings, just data in, data out. Watson won't miss clues on those rare cases because he's simply not prejudiced to rely on the easy answers.

Soon after its *Jeopardy!* triumph, IBM began working with doctors and researchers at Columbia University to develop a version of Watson that won't be a mere novelty in health care but a true caregiver and diagnostic authority. In September 2011, the giant health insurer Well-Point announced plans to give Watson a job assisting doctors in their offices with diagnoses, providing a valuable and legitimate second opinion. WellPoint's main purpose in using Watson is saving money, but in paying IBM for Watson's time, patients also receive the benefit of more correct initial diagnoses.

Herbert Chase, a clinical medicine professor at Columbia, tested Watson with a vexing case from earlier in his career when he had to treat a woman in her midthirties who complained of fleeting strength and limp muscles.[24] The woman's blood tests revealed low phosphate levels and strangely high readings of alkaline phosphatase, an enzyme. Queried with these conditions, Watson ventured that the most likely malady was either hyperparathyroidism or rickets. Crunching the test results did cause Watson to allow, however, that the woman should be checked for an atypical case of rickets that was resistant to treatment with vitamin D, which is exactly what she had.

Getting to that result took Watson a few seconds. For the doctors on the initial case, it took days.

7

CATEGORIZING HUMANKIND

ALL OF US HAVE A personality type. All of us, when observed by a true expert, can be categorized, tagged, and cataloged like an animal in a zoo. Are you a volatile orangutan or a docile fawn? Do you operate with guile and trickery, or are you as honest as Abe? Psychiatrists with elite skills can, just by talking with us, answer these questions with surprising swiftness. They know why we work, they know why we fight, and they know the kinds of people with whom we will get along. Diagnosticians of this caliber don't reside in most psychiatry practices; they're a rare breed. But what if their type could be imitated, reproduced, and even enhanced by a machine? What if there existed a set of algorithms that could identify our character, know our weaknesses, determine our thoughts, and predict our actions? What if we created a machine that could read our minds?

Such a development would change not only psychiatry but also all manner of commerce, customer service, and hiring practices. What if, after a business call, you could consult a bot that told you exactly how the call went and what the other side thought of your proposal? Sales calls would never be the same, bad relationships would end earlier rather than later, and negotiations would be straightforward—until both sides employed unflappable bots to do their talking, that is.

Such technology will disrupt many of our day-to-day interactions with other people. It may sound like a far-off prospect, but algorithms have already figured many of us out, parsing our personalities and learning our motivations. We're rarely given explicit warning when a bot is observing us; it's something that's crept into our lives with little fanfare. While it may sound creepy, an algorithm that can read you as well as your spouse can be useful.

We've almost all been read by a bot, although most of us didn't realize it at the time. You likely recognize these words: "This call may be recorded or monitored for quality and training purposes." Whether it's a bank, credit card, airline, or insurance company, when we dial up customer service and hear that familiar phrase, we've allowed an algorithm to drop in for a listen. It's not unreasonable to believe that this message is merely served in case a supervisor or manager is listening in or may replay the call later. But in many cases, that's not what's happening.

What is happening is that a bot is eavesdropping on your call. It listens to you talk, assesses your personality type, and determines why you've called. In some cases, the bot relays this information to the customer service agent as they talk to you. Most incredibly, the bot tries to read your mind. How can you be most quickly and cheaply satisfied? Once the bot gets to know you—it can figure you out in thirty seconds or less—it will route your future calls to agents who share your personality traits. Pairing callers with the wrong agents heightens the risk of a colossal argument. Putting like-minded customers and agents together results in shorter calls, higher customer satisfaction, and bigger profits for the company.

To better understand how these bots work and the breadth of their abilities, it helps to understand their roots, which trace to NASA and, as usual, Wall Street.

PICKING THE RIGHT PEOPLE: FROM LUCK TO A SCIENCE

On April 11, 1970, three men sat in a 363-foot-tall Saturn V moon rocket filled with superchilled oxygen and hydrogen that, as it burned, would propel them up and out of the earth's atmosphere. From there, they would again light their booster to leave earth's orbit. This was Apollo 13. Destination: the moon.

The next morning the men radioed mission control. As was the custom, the mission control man fed the astronauts the news of the day. This was his message, starting with a baseball score from Houston, where the astronauts and mission control spent most of their time:

> The Astros survived, 8 to 7. . . . They had earthquakes in Manila and other areas of the island of Luzon. West German Chancellor Willy Brandt, who witnessed your launch from the Cape yesterday, and President Nixon will complete a round of talks. The air traffic controllers are still out on strike, but you'll be happy to know the controllers in Mission Control are still on the job.[1]

The capsule crew, James Lovell, Fred Haise, and John Swigert, appreciated the humor. Later in the conversation, a joke about filing taxes (April 15 was approaching) reminded Swigert that in the rush to get ready for the mission, he had forgotten to file. Through their laughter, the mission controllers promised to inquire about a tax extension for him.

During routine midflight maintenance on the spacecraft, a malfunctioning thermostat sent an electrical arc into one of the giant tanks of oxygen, which ignited immediately, sparking an explosion that blew half of the ship into space.

The men were faced with a command ship quickly running out of oxygen. To survive, they retreated to the only spot they could: the lunar module, a vehicle that would never see its intended destination. The lander, designed to keep two people alive for only forty-five hours, would now have to sustain three men for nearly one hundred hours. To conserve enough battery power to survive, the men shut down most of the electronics in the lunar rig, including the heating. As the hours and then days wore on, the temperature approached freezing inside the ship, preventing sleep and sapping the men of energy.

Mission control in Houston guided the men into a maneuver that would swing their craft around the dark side of the moon and then back toward earth, using the moon's gravitational pull as a slingshot. Gingerly nudging the lunar module's thrusters, the men got their ship into perfect alignment for the maneuver. It was a technical masterstroke.

But that trick was only one of many brilliant engineering adaptations conjured by mission control in Houston. At one point, the carbon dioxide levels in the lunar module grew dangerously high as the extra man's exhalation overwhelmed the craft's CO_2 scrubbers, which were made from cartridges of lithium hydroxide. The cartridge supply had run out. Without a fix, the men would suffocate on their own exhaust. The damaged service module had cartridges of its own, but they were of a different shape and wouldn't fit the lunar module. Mission control instructed the men to jury-rig a filter box to fit the other cartridges with items in the ship: a plastic bag, the cover of a flight plan book, and a roll of gray tape. The apparatus was attached to one of the scrubber's oxygen hoses and, miraculously, worked. The CO_2 dived back to nonlethal levels.[2]

That the men would splash down in the South Pacific two days later is a testament to the capabilities of the team in Houston as well as that of the astronauts themselves. Just as remarkable, however, was the resiliency and cohesion of the three-man crew under pressure. For one hundred hours, these men sat inside a tiny capsule designed for two, constantly performing one critical task after another. When subjected to such dire conditions, individuals react in acutely different ways. Some people tuck their emotions away and calmly go about the task at hand, no matter how foreboding the immediate future may appear. A good chunk of us, however, even astronauts, are unable to escape paralyzing fear; we recede, we do nothing, or we act irrationally when faced with the likely prospect of death.

It's difficult to predict how a person will react. Even harder to forecast is how well people will get along and function *together* when put into such isolated conditions. With Apollo 13, NASA got a little lucky. One of the crew members, Swigert, had only joined the mission a day before launch. He was a replacement for Ken Mattingly, who had been exposed to measles the previous week. The trio had little time to get used to working together before they were strapped into seats and blasted into outer space. But NASA did have some idea of how the three would mesh. The space program had been assessing the compatibility of astronauts' personalities for nearly a decade. This psyche-measuring program played a crucial role in guiding the Americans ahead of the Soviets in the space race.

Without thorough and exhaustive psychological assessments, there was no way to know how a three-man cocktail would mix in a tiny capsule, not to mention during missions as taxing as a lunar landing. NASA was more than aware that a bad match of astronauts could compromise a mission and even the lives of the crew. That's one of the reasons why almost all of the early astronauts were plucked directly from the military, where the candidates had been exposed to combat, stress, and death. Peer and background checks on a candidate's character under fire were paramount to the selection process.

Even so, after the close call of Apollo 13, NASA wanted a better method, something that would be rigid, absolute, and accurate. So in 1971 the agency set out to create a comprehensive system of categorizing people to the point of knowing who would work well together and who was liable to explode if exposed to the wrong mix of pressure and people. NASA needed a way to read people's minds and predict their exact actions. The system the agency developed in the 1970s and 1980s lives on today in the form of bots that read our personalities, our wants, and our intentions—a fact of which most of us are oblivious.

NASA developed its people-assessing algorithms in the focused time of the Cold War, when creating better astronaut crews than the Soviets, who experienced at least two compromised missions due to crew conflict, was all that mattered. NASA didn't have high-powered computers and speech-recognizing software, so personality assessments were carried out on paper. These methods were carefully scripted by NASA's head psychiatrist at the time, Terry McGuire.

As a child, McGuire found that he had a talent for being a peacemaker between bullies and their introverted targets. "I knew as I was growing up that I was more sensitive than most of the guys to emotional things going on around me," he remembers. He could innately sense suffering, panic, fear, and anger boiling in people before the clear signals had manifested. McGuire became so interested in people and their behavior, in fact, that he took up rock climbing and mountaineering not only because he enjoyed the sport but also because he was able to observe different people's reactions to fear, stress, pain, and adversity. "It was an incredible laboratory," he says.

In 1958, the air force enlisted McGuire to develop systems that could predict soldiers' reactions under life-threatening stress. To this day, the Pentagon runs dozens, if not hundreds, of top-secret missions every year. Not all of them are as high profile as the raid that killed Osama bin Laden, but they all require men who don't flinch under fire. Some men who otherwise seem to have nerves of steel utterly fail in such circumstances. And others, sometimes the least likely of troops, turn emotionless, focused, and deadly. The military wanted to know, why do some guys do so much better than others? And, even more important, how do we know who they are ahead of time?

McGuire was to be the self-described "psychiatrist on the dark side of the air force." The laboratories, instruments, and resources at his disposal were superior to any academic or hospital setting at the time.

It was the Cold War; the military was the cutting edge for the world's technology, and now it was at McGuire's fingertips—all so he could follow a subject of curiosity that had been with him since adolescence.

McGuire dug in, and eventually developed a three-hour set of questions that he would deliver to new recruits. From their answers, he stood a better than 50 percent chance—as much as anybody had hoped for—of knowing a man's makeup when he was placed under extreme circumstances. McGuire helped send team after team into dangerous missions, their purposes forever classified.

McGuire pitched to his superiors the idea of tagging along on a mission himself. He needed experience in the field, he reasoned, to know what to look for in these men. At some point, permission was granted and McGuire found himself in the back of a helicopter, flying low over the jungle in Vietnam. A small group of American soldiers had been pinned down on a mountaintop. They were running out of ammo; when their last shell hit the ground, they would be overwhelmed by teeming enemy forces.

The resupply team flew in the middle of the night, their lights off and the chopper blast whipping their faces through the open doors. As the helicopter approached the mountain, hundreds of tracer bullets zipped out of the jungle, pelting the craft's metal hull. McGuire and the others aboard didn't fire back; they didn't want to betray their position with muzzle blasts. It took incredible restraint, McGuire recalls. They landed in the only open patch of ground on the mountain and dumped a big load of ammunition and explosives. "And then we got the hell out of there," McGuire says.

Through his investigations in the field, McGuire found indications of what kind of people thrive under pressure, but he still found the practice less than a science. At times, he still felt he was just guessing. Sometimes he was straight wrong. In 1971, his research attracted an interested party in NASA, which was beginning to introduce scientists into the astronaut program and it wanted to know how to pick them. It wanted to grade its candidate pool and figure out which of its astronauts would work best together. The agency had previously plucked most of its astronauts from the air force's test pilot program. These were the steely men who flew the country's newest, least known, and most dangerous aircraft.

The test pilots were unlikely to crack while in space. As for the scientists, well, "NASA just had no idea," McGuire says. NASA brought him on board because it wanted precision—and it wanted McGuire to vet a pool of more than five thousand scientists. It needed to know not only who would hold up under pressure but also how to determine people's compatibilities. Sometimes stress couldn't break people, but combined with friction with peers it could lead to worse fractures.

"They were asking me to do things that people really hadn't done," McGuire says. "They wanted me to predict how candidates would respond if the hatch started to squeal when they were 150 miles up."

McGuire had his own methods, but he started combing through the others' research to see if anybody else had taken an approach he hadn't thought of. When he found something interesting, he often invited the psychiatrist or researcher to NASA in Houston to evaluate candidates with him and compare notes. At one point he came across the research of Taibi Kahler, a psychologist at Purdue University.

In 1971, Kahler developed a process to identify different kinds of human communication and their efficacy. It narrowed verbal interactions down to a number of sequences that were identified by speech patterns and word choice. Each sequence could be objectively identified as it tumbled out of the speaker's mouth. Different combinations of sequences signaled whether the person was engaged, happy, bored, disgusted, or repulsed by their conversation partner and the subject. With Kahler's system, there was no looking between the cracks or trying to peer into a subject's soul. All of the information needed to assess a person came out of his or her mouth.

Kahler built his system around the words that we speak. Our sentence structures, our chosen verbs, predicates, and pronouns all betray well-defined characteristics that most people have carried with them since childhood. Putting together sentences is something we do hundreds or thousands of times per day. The process becomes incredibly ingrained for us; we cannot, with the flip of a switch, break these habits or pick up completely different ways of speaking. To somebody with the skills of Kahler, we can't disguise who we are.

Kahler's linear method of dividing people up according to clear patterns of speech made it a natural fit to be programmed into an algorithm and, eventually, a bot. Psychiatry has long been thought of as a field that would seem impervious to technological advances. But Kahler's method—using the theory of logic trees and binary determination—may have proven this to be untrue.

In 1972 McGuire didn't think Kahler was right, but he wanted to see the professor's methods for himself. So McGuire invited Kahler down to NASA; it couldn't hurt—dozens of psyche-evaluating doctors passed through every year. Kahler sat in with McGuire and his team as they interviewed and judged astronaut candidates, who were run through an hour or two of standard queries before being subjected to an intense string of questions and tests for which they couldn't possibly have prepared. Succeeding in this gauntlet represented a life's goal to most of the candidates. It meant they could withstand the pressure NASA required—they wouldn't break, like so many other candidates.

The NASA people furiously scribbled notes during the candidates' sessions, which could last for several hours. Kahler, tasked with making the same assessments, would carefully observe the subjects while taking his own notes in tidy fashion on a legal pad. McGuire noticed that after ten minutes or so, Kahler would abruptly put his pencil down. He'd seen all he needed to see. He did this for every single candidate. The others would carry on for hours, but Kahler just sat calmly, his knees crossed, his hands folded, and his pad and pencil motionless.

Kahler's methods seemed hasty and McGuire guessed they would prove ineffective. But as McGuire's team followed candidates through training and their interactions with others, Kahler's predictions held true, including anticipated clashes among trainees. McGuire's assessments, as well as those of everyone else at NASA, reverted to the mean (some right, some wrong), but the doctor from Purdue was consistently spot on. Somehow, Kahler's pithy evaluations were more effective than anything else McGuire, a pioneer in this field, had seen. Kahler could diagnose a person's character, fiber, and future actions by observing them for fifteen minutes.

Kahler's methods—and more important, his accurate results—excited McGuire and NASA. But could he teach others to make the same observations he did? The answer was yes. Thus began a NASA project that would determine the makeup of the space-bound crews representing the United States of America. The science—and it can be safely called that—put the right scientists on missions and avoided human-made catastrophe: mutinies in the void of outer space.

Mutinies were a possibility. NASA regularly listened in on conversations between Soviet cosmonauts and their ground controllers. On one particular mission in the mid-1970s, Kahler recalls, it became clear that the two men in the capsule weren't getting along. The men's daily space tasks included taking blood samples from their partner. Halfway into the trip, one of the Russians purposefully drove the needle so hard into the other's hand that it pierced bone. A scuffle erupted and Russian ground control operators worried about the possibility of a space homicide. Nobody died, but the afflicted man's hand became hopelessly infected, swelling into a giant pus-filled ball. To preserve his life, the mission was aborted and the men returned to earth.

McGuire says such disasters were closer to happening on many NASA missions than people realize. To avoid them, he became a rapt student of Kahler's methods. The professor's system—and the slightly modified version with which eavesdropping bots evaluate you when you call—classifies people under six categories:

1. **Emotions-driven people.** They try to form relationships and learn about the person they're speaking with before diving

into the issue at hand. Women comprise three-quarters of this group, which makes up 30 percent of the total population. Tight situations make this group dramatic and overreactive.

2. Thoughts-based people. This group tends to do away with pleasantries and go straight for the facts. A rigid form of pragmatism drives most of the decisions they make. Under pressure, they tend to become humorless, pedantic, and controlling.

3. Actions-driven people. Most used-car salesmen would come under this umbrella. They crave progress and action, even in tiny chunks. They're always pushing, prodding, and looking for an angle. Many in this group can be charming. Pressure can drive actions-driven people to irrational, impulsive, and vengeful behavior.

4. Reflections-driven people. This group can be calm and imaginative. They often think more about what could be rather than working with what already is. These people, when they're interested in something, can drift off for hours while they dig into the new subject. Applying this knowledge to a real-world project, however, is a weakness.

5. Opinions-driven people. The language used by this group is stocked with imperatives and absolutes. They tend to see one side of a situation and will stick to their views even when refuted with proof. More than 70 percent of politicians are opinions-driven people, but the group constitutes only 10 percent of the population as a whole. People from this group can also be tireless workers, who will persistently grind away on a problem until it is solved. Under pressure, their opinions can become weaknesses; they can grow judgmental, suspicious, and sensitive.

6. Reactions-based people. Kahler called this group *rebels,* but the modern set of standards—and those on which personality-deciphering bots are built—refer to them as reactions people. This group is spontaneous, creative, and playful. They react strongly to things: "I love that!" or "That sucks!" Many innovators come from this group. Under pressure, they can be stubborn, negative, and blameful.

If you think you spot shades of yourself in more than one of the categories above, you're not mistaken. Our personalities tend to be dominated by one of the six traits, but we possess differing percentages of the other five. Donald Trump, for instance, is most certainly an actions-based person, but he likely has an emotions-driven piece of his personality that isn't part of his public face.

As one of NASA's powerful people determining who would be part of its manned space program, McGuire became an expert at quickly reading people. He can tell within five minutes what personality is dominant in somebody he's talking with. Strong thoughts-driven people issue a lot of questions; they're always gathering data. Opinions-driven people issue edicts, see things in black and white. They start many statements with an "I think," later treating their "thoughts" as facts. Reactions-based people like leading dialogues with a "Yes!" Over time, McGuire came to learn what kinds of personalities were likely to butt heads in space and which ones were simply unsuited for the astronaut world. Reflections-driven people, for instance, aren't a good fit at NASA. "They're virtually nonexistent in the program," McGuire says.

People have varying degrees of compatibility. All of us have met that insufferable clerk at the bakery or the ornery barista. But it's also true that other people may not be so annoyed by the same person. That's because their personalities offer a closer match. It would be unwise for NASA to stuff an emotions-driven person into craft with three largely thoughts-based people. Or even choosing four strong thoughts-based candidates could make for trouble. It's best to mix in others—like a reactions-based person and an opinions-driven person. The leader of this group may well be a thoughts-based person, but McGuire would see to it that his personality included his crew's primary traits; having those components would allow him to understand his astronauts and better manage relations.

Some of the most effective commanders in space are those with well-balanced qualities. While many pilots tend to be thoughts-driven, the best ones also have good pieces of emotional- and opinion-driven personalities. People don't like reporting to pedantic workaholics with no compassion, but they will fall in line under those who work hard and display genuine empathy for their subordinates' jobs and lives.

"John Glenn, for instance, who is steady as a rock, oh man—to see him moved to tears by the misfortunes of another person—that shows you that sensitivity and toughness are independently determined," explains McGuire.

NASA also wanted steady hands on the ground in mission control. Apollo 13 would never had made its unlikely journey back to earth without resourceful and unflappable people sitting at the screens in Houston. McGuire oversaw the psychiatric evaluations of both the space-bound men and those who, chain-smoking from their swivel chairs on earth, helped guide them to safety.

When McGuire saw potential eruptions between crewmates, he learned how to take action ahead of time. He'd often sit down with the astronauts he was worried about and tell them candidly what he saw coming. The honesty worked wonders as both crew members would go out of their way to keep their relations civil and professional; nobody wanted to jeopardize their standing in the program. McGuire would also alert the commander of the mission to be on the lookout for potential problems and tell him how to disarm ugly conflicts in space.

Of the several dozen operations he saw take off, McGuire predicted crew trouble on six of them. For five of those missions, he was right. He recalls catching an astronaut named John in the hallway following a press conference after his crew had returned to Houston. McGuire had been concerned about John and one of his fellow shuttle mates setting off a space melee.

"So, how'd it go up there?" McGuire asked.

"Oh, it was good," replied John affably. Then his gaze narrowed and his eyebrows creased. “But one more day—just one more day—and I’d have killed him.”

McGuire believed it.

OUT OF NASA AND INTO OUR LIVES

As other people in the psychiatry and human behavior methods learned of NASA’s success, Kahler’s methods spread. Soon Fortune 500 CEOs sought out McGuire and Kahler for tutoring on how they could lead more effectively and connect with their employees. Kahler even prepped Bill Clinton for his speeches and debates during the 1992 presidential race, coaching the future president on what words and phrases to avoid when courting individuals of varying personalities as well as how to handle addressing a wide, moderate electorate. Clinton may be a natural communicator, but even he found the exercise useful.

A clip from one presidential debate in October 1992 clearly shows how Kahler’s methods affected Clinton’s delivery. In the town hall format, George H. W. Bush and Clinton both answer a question from a distraught unemployed woman who wants a stronger economy and more jobs. She is emotionally upset. Bush replies with a spirited recitation of initiatives. Clinton, though, walks over to the woman and talks about how he feels the pain of people who have been laid off in Arkansas, how he knows the plant managers by name, knows the jobs they used to do, the names of their children, and why they work so hard. Clinton says he wants to invest in America, in its people—a vague measure, but one that stirs the woman, the crowd, and the TV audience alike. He met her emotional personality with empathy rather than the cold calculus of a thoughts-based person as Bush had done.

About the time Clinton was running for president, one of Kahler’s disciples, Hedges Capers, a psychiatrist, started consulting for Telecom Technologies, a hardware maker that wanted better relations between its C-level executives and the cubicle-bound folk. Today Capers is one of the best readers of people in the world; he thoroughly knows a person's personality after a thirty-minute chat.[3] The O. J. Simpson defense team asked him to assess jurors' personalities in its 1995 trial. Working with Telecom, Capers became close with Kelly Conway, who eventually rose to be CEO. The few years the two worked together were formative ones for Conway. He became engrossed with the theories and methods established at NASA for evaluating, reading, and predicting people.

Conway eventually left Telecom to found eLoyalty, a consultancy for companies with large call centers, and he buried away in his head what he'd learned about assessing people. "I had a feeling it would someday become really useful," Conway says.

Conway's eLoyalty grew and did well; it eventually went public and its stock was traded on the Nasdaq. But Conway didn't forget about NASA or Capers.[4]

People's core personalities rarely change. Exceptions happen when a person is exposed to long-term abuse and stress, but these cases are outliers. As we get older, our vocabulary may grow varied and dense, but the way in which we speak—the structure we lend to our sentences—remains the same. Those structures give away amazing clues as to how any of us process logic and act under pressure. By exploiting this fact, NASA had remotely hacked the human brain.

In 2000, Conway's eLoyalty started losing business to larger consultancies with more expertise in moving some call center personnel offshore. As he looked for other ways to grow eLoyalty, Conway found few paths not already clogged with competitors. The business of call center software offered only small paths of growth that would have to be won inch by inch. The only way to leap far ahead of the pack would be to innovate, to create something that no other player could offer.

The seed of NASA's work had begun to bloom in Conway's head. Along with the working of the space agency's process, Conway knew two other things well: his own business of call center optimization and, as CEO of a Nasdaq company, the latest and greatest tricks being performed on Wall Street.

Knowing the strides made in speech recognition, in 2000 Conway thought that perhaps a mind-reading bot could be constructed after all. His first phone call went to Capers. The psychiatrist considered Conway's pitch and thought not only that it had merit but also that it could change the world of business and the practice of psychology.

Inventing a product to change customer service could by itself prove immensely lucrative. The customer service industry is far larger than most people realize. AT&T, for instance, has 100,000 seats in its call centers and spends $4 billion a year to run them. The four million call center employees in the United States, in fact, represent the third-largest occupational category in the country. Inventing a better tool in this industry could be worth multiple billions of dollars every year.

For years, constructing a bot that could quantify spoken words and determine personalities and thoughts was impossible. The technology—software and hardware—just wasn't ready. Speech recognition software—the ability of computers to capture and translate exactly what humans say—was a lost cause for decades. The software that did exist for the purpose was buggy and often wildly inaccurate. But in the early 1990s, two scientists at IBM's research center dove into computerized speech recognition and translation, a field that had long failed to produce anything robust enough to be used in everyday situations. Peter Brown and Robert Mercer started by working on programs that translated one language to another, starting with French to English. Most hackers working on the problem up to that point knew both languages and wrote programs that translated words directly: the English *ham* is, in French, *jambon; cheese* is, of course, *fromage;* and so on. But all languages are rife with exceptions, strange rules, and counterintuitive idioms and phrase that severely complicate writing translation algorithms. There are simply too many inconsistencies to account for.[5]

Brown and Mercer didn't know French, nor did they care to learn it. Instead of writing thousands of algorithms by hand, they fed transcripts of Canadian parliament meetings into their IBM workstations. Because Canada has two official languages, the transcripts, which for a single year can number in the millions of words, are available in both English and French. The men created machine-learning algorithms to look for patterns in the twin texts. Where others had tried to solve the problem with elegant code that attempted to reproduce the grammatical structures of different languages, Brown and Mercer employed "dumb" software and brute force. "Only in this way can one hope to hear the quiet call of (marqué d'un asterisque/starred) or the whisper of (qui s'est fait bousculer/embattled)," wrote Brown in a paper summarizing their research.[6]

Brown and Mercer then built a set of algorithms that tried to anticipate what words would come next based on what preceded it. Their hack was so revolutionary, in fact, that it not only changed speech translation software, but also speech recognition programs. Instead of trying to nail each word as it comes out of the speaker's mouth, the latest and best speech recognition software looks for strings of words that make sense together. That way, it has an easy time distinguishing *are* from *our*. *Are you going to the mall today* won't be mistaken with *Our you going to the mall today* because, simply, people never say *our you going*. Just as we learn grammar rules, so the machine-learning algorithm did as well. This method forms the backbone of the speech recognition programs we use today.

Brown and Mercer's breakthrough didn't go unnoticed on Wall Street. They left IBM in 1993 for Renaissance Technologies, the hedge fund. Their work developing language algorithms could also be used to predict short-term trends in the financial markets, and versions of their algorithms became the core of Renaissance's best funds. During a run powered by Brown and Mercer's work, Renassiance went from $200 million in assets in 1993 to $4 billion in 2001.[7] How the speech recognition algorithms are used in the markets isn't exactly known, which is why Renaissance remains so successful. Renaissance is the largest enigma on Wall Street—and its knack for luring top science talent to finance is second to nobody, including Goldman. When James Simons, Renaissance's MIT-educated mathematician founder, retired in 2009 after becoming a billionaire with the help of Mercer and Brown, he named them co-CEOs.

The work of Mercer and Brown meant that by 2001, Conway could buy off-the-shelf software that could transcribe most conversations with a reasonable degree of accuracy. Now that he could capture people's words, Conway needed to build algorithmic bots that could take those transcribed words and apply to them the calculus of NASA.

To do this, Conway set out to develop a library of algorithms for every series of words or pattern of speech imaginable. Anything a person says can be read for personality clues. For instance, if a person opens the conversation with a customer service representative by asking questions and demanding answers, dispensing with pleasantries and chitchat, it's likely they're a thoughts-based person. They may say things like, *Hi, my car needs a new muffler. What is your repair price for a 2007 Subaru Outback and when can you do the work?* The keywords in this statement are *needs, what, when*—the caller is demanding information and he wants an answer as quickly as possible.

Another caller might call and explain how the poor performance of the company's product made her feel. This person might say, *I'm exasperated—this doesn't work and I'm just not sure what to do anymore. I expected more out of your company and your service. I'm disappointed.* The keywords here are *exasperated, disappointed,* and *expected.* Also, the caller has further indicated that she is an emotions-based person by heavy use of the personal pronoun *I,* followed by declarative statements of feeling.

The first caller was on a fact-finding mission. Meet his answers with facts and he will be quickly satisfied. Making small talk, the algorithms should know, would annoy him. The second caller wants a heavy dose of empathy and general conversation. She will be best satisfied by an apology and genuine understanding. Conway needed algorithms that would pick up on these facts in a matter of seconds.

Going past personalities, Conway wanted to build a bot that would know what callers wanted before they even asked for it. If a person threatened legal action, for instance, Conway wanted a supervisor to know immediately, especially for potential clients such as health insurers where litigation was a frequent enough phenomenon that it affected the bottom line every year. Any kind of automatic measure that could head off just a handful of lawsuits would be worth millions. Conway knew that handling callers the exact right way on a consistent basis could lift a company's results and build a loyal customer base.

Knowing that creating a revolutionary product could result in a huge payoff, Conway looked to change the direction of his public company based on a hunch and the belief that something could be built to read people's minds. Not everybody was convinced. Conway told Robert Wert, eLoyalty's head lawyer, that he knew of a psychological model that could predict human behavior. Wert nearly guffawed. "I said, 'Right! I've been a lawyer for twenty years and, yeah, I can predict humans too—they're all assholes.'"

But Wert was converted when he met Capers and saw the science in action. The lawyer now runs Kahler's consultancy. Law firms pay Wert more than $10,000 a day to tutor their partners on how to decipher humans quantitatively.

Conway wanted to take humans out of it. Because it was a quantitative method, it was possible, Conway figured, to build bots that could read customers' minds. "We are categorizing the human language," he says.

With the small team he had, Conway kept working on his algorithm library, building it up to go live. At the same time, he set out with Capers to find out if his theories would even work. He had to know what the real advantages were of knowing a person's thoughts, of knowing their motivations and whether this whole mission of his would prove useful—providing he could even pull it off.

Conway talked Vodafone, the global wireless carrier, into turning over to him fifteen hundred phone call recordings for a group of twelve customer service agents. Having more than a hundred calls per agent allowed Conway to determine how each of them fared with all different types of callers, from actions-based people to reflections-based people. He could also figure out what kind of personality each of the Vodafone representatives possessed and therefore test his hypothesis that like personalities reach resolutions faster and leave fewer callers hanging up angry.

Conway and Capers listened to each call independently and classified the personalities of all fifteen hundred callers and the agents. They then timed each call and classified it as a happy conclusion or, in the worst case, a potential customer lost. The pair then put all of this data into a matrix and examined the results.

The results were nothing short of astounding. When a caller reached a personality much like their own, calls lasted about five minutes and the problem was resolved 92 percent of the time. That's high-quality customer service by any measure. When a customer's call landed with an agent with an opposite personality, however, the difference was amazing: ten-minute calls and a 47 percent problem resolution rate. This meant that calls matching the personality of the caller with that of the agent were twice as effective as other calls. With drastically shorter call times, a company like AT&T could eventually eliminate a third of its call center costs—provided that Conway could actually build the kinds of bots necessary to make it possible.

As it turned out, it was very possible. To make it reality, though, would prove to be a massive project. To build the algorithms necessary, Conway hired crack programmers along with "just the smartest people I could find." He found that hiring the brightest talent and then training them on what he needed done was often more effective than simply hiring people who already had the correct programming and logic skills. It's akin to picking Michael Jordan for your flag football team over somebody who may have been an all-conference wide receiver in high school. Sure, the second guy can catch and knows how to play football, but MJ's natural athleticism will likely blow by him once he gets a feeling for the game.

To create a mind-reading bot, Conway needed a lot of Michael Jordans. He was prepared to take all of the profits from eLoyalty's core business of call center management and plow them into this project, something he felt could change the world far beyond customer service. With tens of millions of dollars in operating profits per year, Conway could afford to be generous in hiring his team. But when it came to actually signing a large team of superstars, he found that other well-financed suitors were wooing them too. The other suitors were, in fact, as well funded as any in the world. Conway had run right up against the Wall Street quant machine.

8

WALL STREET VERSUS SILICON VALLEY

A COLLECTION OF EVENTS, ALL coalescing right around 2001, led to the fattening of trading houses' income statements and the pools of adroit programmers they kept on staff. The expansion of the financial sector beginning in this period was unprecedented and fueled by two things: the booming housing sector (which Wall Street helped create) and the rise of algorithmic proprietary trading.

In 2000, bots accounted for less than 10 percent of all trading on U.S. stock markets. Big Wall Street players knew well of their existence, but the algorithms back then weren't moving the market. The thought of them co-opting the entire works and sparking an event like that of the Flash Crash was inconceivable. In 2000, people controlled the pits, not computers. Almost all trades on the NYSE still traveled through the hands of a human specialist, who had the unique privilege of overseeing most of the trading on one particular equity. On the floor, traders could shoot orders only as quickly as the specialists and their minions could handle them. The system was entirely human. It was still the pits, not computer motherboards or the pulsing light in a fiber-optic line, that ran Wall Street. Peterffy's system and a generation of copycats had taken over the Nasdaq, but the NYSE remained a place where yelling out orders trumped typing, let alone anything wholly automatic.

All of that was changing, however. The rise of the electronic trading networks made it easier and cheaper for trader/programmers like Peterffy to wade into markets once controlled by cliques and a select few. A solid network connection coupled with a powerful desktop computer and some well-written code could spring a programmer with market knowledge straight into the game.

In addition to greater access, the so-called decimalization of Wall Street gave hackers another huge opportunity to make money. Before decimalization, U.S. stock markets had always operated on fractions of a dollar. For instance, before 1997 the smallest increment a stock could move had been one-eighth, which is equal to 12.5 cents. The increment was then halved to one-sixteenth, but that still meant the minimum spread on a stock was at least 6.25 cents. This was still a large enough spread to keep the established brokers and market makers, even those who had underinvested in technology, raking in money. All they had to do was sell at the offer and buy at the bid. The use of fractions guaranteed brokers profits at the expense of everybody else, especially small investors.

In 2001, the Securities and Exchange Commission mandated that all markets switch to decimals, rendering the spreads in many heavy-volume stocks to one penny. Now just being a market maker with clients wasn't enough to cinch up big earnings. You needed volume, lots of it. The only way to get it: technology and the brains that knew how to use it.

By early 2008, automated bots accounted for 60 percent of all U.S. stock market trades, and the financial industry had spent seven years sucking up every deft graduating engineer, physicist, and general Renaissance man who had even a mild attraction to a large starting salary and a bonus big enough to buy most Americans' homes twice over. Wall Street had grown to become a larger hirer of math, engineering, and science graduates than the semiconductor industry, Big Pharma, or the telecommunications business. It was into this headwind that Conway tried to hire for the better part of a decade.

Conway's eLoyalty, based north of Chicago in Lake Forest, spent dozens of days recruiting at nearby Northwestern University and the University of Chicago. Conway would show up at recruiting events with a few of his most trusted employees and set up his company booth. But as Conway and eLoyalty dug further into the recruiting game, they found themselves stymied by Wall Street's recruiters. Goldman, Morgan Stanley, Citi, Credit Suisse, and the rest of their lot were omnipresent at top universities' career events. In most cases, the Wall Street firms sent junior employees who had only been out of school a few years to such events. They could talk shallowly about what working at their companies was like. They didn't pitch vision, they didn't pitch challenge, they didn't pitch intellectual stimulation—they didn't really pitch. All Goldman or Morgan had to do was show up and their booths would be mobbed.

This period—2001 to 2008—will likely go down as Wall Street's pinnacle. The firms enjoyed not only record revenues, profits, and head count, but also widespread prestige and acceptance on a level that they had never seen before and will likely never see again. This was what Conway was up against.

Conway would get a few people trolling past, but getting a crack at the best people proved difficult. "It's just really hard to beat out a Goldman Sachs when the candidate has never heard of your company," he says. Even when he could get a candidate interested, he'd get outspent by trading houses' deep pockets. "We pay well, but we can't throw $200,000 at a fresh college graduate," he reasons.

Conway did everything possible to land the best minds. When he thought the company's suburban headquarters in tony Lake Forest might be a turnoff to graduates looking to stay in the city, eLoyalty opened up a large office in the heart of Chicago's downtown, right in the financial district where rents are highest. He rewarded employees with cash for hiring referrals. Conway kept pounding at new graduates, hoping to find a stream of bright, like-minded people who might start a trend by heading to eLoyalty.

Conway tried to short-circuit the process, bringing in MBA graduates from top schools with experience in linguistics. Surely they could conceive the algorithms he needed. But that route didn't work. "We overpaid and it turned out they weren't very good at their jobs," Conway says. The MBAs got caught up on details, procedure, and metrics. What Conway really needed was technical people charged with creativity—high-capacity minds that could quickly conceive of problems and find unique and disparate ways of solving them.

A typical (and large) problem facing eLoyalty: Not every emotions-based person uses the exact same patterns of speech. Devising ways to automatically sniff out these differences and still categorize people correctly requires the creation of tens of thousands of algorithms capable of spotting all emotions-based people and their quirks, intentions, and language. Only people with a supreme capacity for focus and a highly analytical mind are up to the task.

Conway did make some progress on hiring these kinds of people, albeit slowly. Prospects were given personality and IQ tests. Only the best scorers were considered. More than a quarter of all of eLoyalty's hires are in the top 1 percent of IQs nationally. This is exactly the same candidate so cherished by Wall Street. It's these people who can unravel the twelve things affecting a securities price and how to conceive of algorithms to trade on these theories at the speed of light. Competing against a blue-chip name like Goldman Sachs, plus salary/bonus compensation of $400,000 for even junior-level people, slowed Conway's progress.

Conway wasn't alone in his losing battle to wrest talent from the temptress of Wall Street. The entire tech world watched as investment banks, trading houses, and now a new breed of company, the boutique high-speed trader, all swarmed elite universities in search of talent. The changing dynamic on campuses wasn't a small one. For instance, the share of MIT graduates heading to Wall Street popped 67 percent from 2000 to 2006, when more than a quarter of all of the school's graduates headed into finance.[1]

Conway and the rest of the tech crowd also had to battle a drastically shrinking talent pool. The field had expanded in the 1980s, when, following the rise of the personal computer, students poured into computer science as a major. CS departments swelled and programming talent flowed to all corners of the economy. But that period ended abruptly with the tech crash of 1988, which was partially brought on by the 1987 stock market disaster. Following that, computer science remained an unpopular college major for almost ten years.

But then came AOL, Netscape, eBay, and Yahoo! A new class of millionaire was born and computer science schools' enrollments jumped again. When I was an engineering student at the University of Illinois in the late 1990s, the school was riding a wave of alumni successes, including those of Marc Andreessen (Netscape), Max Levchin (PayPal), and later the founders of YouTube and Yelp! The computer science building at U of I was the busiest place on the engineering quad. Students regularly departed for Silicon Valley before they graduated, and stories of dot-com gold kept the quad buzzing. Even mediocre students seemed to be snaring signing bonuses, stock options, and some fuzzy guarantee of Web glory upon graduation.

This story unfolded at a handful of campuses across the country and burgeoned larger as the tech boom grew hotter and its effects reached the whole of America. As a result, the number of college freshmen choosing computer science as a major began to rise quickly in 1998, when computer science majors comprised just 1.5 percent of incoming freshmen. In 2002, the number peaked at 3.5 percent, more than doubling in just five years. But the dot-com crash quickly shrank the pool. By the end of 2007, the numbers had sunk below 1.5 percent. So Wall Street continued to hire more programmers even with fewer of them available overall—compounding the shortage for people like Conway and eLoyalty. The trend became so pronounced that professors started to notice.

Vivek Wadhwa joined the faculty at Duke University to, as he describes it, teach engineering students about solving the real world's problems and to work as engineers—not algorithm jockeys on Wall Street. He was shocked and upset when the majority of his students became investment bankers or management consultants after they graduated. "Hardly any of my students actually became engineers," he says. "And why would they, when they had huge student loans, and Goldman Sachs was offering them twice as much as engineering companies did?"

Wadhwa was once an engineer on Wall Street himself, working for Credit Suisse First Boston in the late 1980s. He eventually became vice president of the bank's information services. Before Credit Suisse, Wadhwa worked at one of Silicon Valley's original innovators, Xerox, where he made $47,000 a year. His first year with CSFB, in 1986, he made $160,000. He made $180,000 in his second year, $240,000 in his third year, and $300,000 in his fourth year. This was all before 1990.

Like many of his own students, Wadhwa figured he'd spend a few years on Wall Street, pay off some debts, and then move on. But he found it hard to pull away. "It gets to be kind of like organized crime," he says. "You can never leave."

Wadhwa eventually got away and founded Relativity Technologies, which made software to help companies migrate from older code bases to newer ones such as C++ and Java. He has since become one of the leading voices of tech education from positions not only at Duke but also at Emory and Stanford.

As a new faculty member at Duke, Wadhwa watched as many of his brightest students ended up on Wall Street, conjuring up the very instruments that would lead the world to the brink of economic collapse—collateralized debt obligations, the Gaussian copula (a fine formula that was misused by the Street), and trading algorithms that could go wild at any moment.

This was in 2007, the all-time height of the stock market. Financial-sector companies were pulling in cash like a vacuum sucks dust. To ensure their spot at the top of the heap, the finance firms needed two things: friends in Washington and the best quantitative brains money could buy. So in both cases they bought them. Going up against this reality, Conway had little chance of expanding with the speed he desired.

BOOM AND BUST OF THE ALGORITHMIC TALENT WAR

The competition for the best algorithm builders extended into and between Wall Street firms. People who could write the most elegant code or conceive of the most clever algorithms could easily shuttle back and forth between players such as Deutsche Bank and Knight Trading. Among this group there exists a level of superplayer, a quant whose skill is so unique and valuable as to command compensation well in excess of $1 million per year.

In some cases, compensation can grow much larger. Mikhail Malyshev, a plasma physicist and the brains behind the high-frequency trading operation at Ken Griffin's Citadel, the quant-centric Chicago hedge fund, was paid $150 million by Citadel in 2008. A year later Malyshev left to start up a competing firm he named Teza Technologies. Assuming he was being robbed of proprietary code, Griffin sued Malyshev, eventually winning a big fine and largely scuttling Teza's business.[2]

Employers like Citadel, while they pay incredible sums to their quantitative minds, are left vulnerable to defections. Ken Griffin's firm had another one of its former financial engineers, Yihao Pu, arrested in late 2011 for allegedly pilfering code before he left. Pu responded by flinging his computer into a river outside of Chicago. Divers recovered the hardware; it contained building blocks of Citadel's most important trading code.[3]

The decadent pay and hiring wars continued until, during the course of one fall day in 2008, everything changed for Wall Street. On

September 15, Lehman Brothers, the international white-shoe investment bank—and a place that employed legions of quants—filed for bankruptcy. The Dow dropped 3,000 points in the next three weeks and most of Wall Street's elite firms teetered near insolvency. Merrill Lynch sold itself to Bank of America, Goldman Sachs secured $5 billion from Warren Buffett, and Bear Stearns disappeared forever.

In less than a year—sparked by the bankruptcy of one company, Lehman Brothers—the vector of the U.S. economy was changed forever. Before this event, the financial sector had been rising as a percentage of GDP at an accelerating rate. Even during the 1960s, a time in which the stock market romped on a long bull run and when Wall Street jobs began gaining a smidge of glamour, the financial sector accounted for less than 4 percent of GDP. In 1982, when quants and programmers first began migrating to Wall Street, not a single bank or financial company had infiltrated the Dow Jones Industrial Average. Not one. But that's when Wall Street went on a tremendous roll. By 2008, just before the collapse of Lehman, the financial sector accounted for more than 8 percent of the economy—a number defying all averages and history.[4] By then, the Dow Jones contained five financial companies, one of which, AIG, commanded a government bailout of $150 billion—bigger than the GDPs of more than two hundred countries, including New Zealand, Pakistan, Kuwait, and Ukraine.[5]

Layoffs hit Wall Street like a virus. Even the seemingly impervious legions of quants, the precious employees who banks and trading houses had spent so much effort and money acquiring, left their offices with cardboard boxes under their arms. For much of the fall and well into 2009, there was genuine uncertainty as to whether Wall Street's biggest hirers of math, science, and engineering majors would survive.

In the fall of 2008, when Conway and his crew from eLoyalty made their normal appearance at a University of Chicago recruiting event, he immediately noticed something different. "They were gone," he says. Goldman, Morgan, all of them. They just disappeared. These companies had not only stopped hiring, they had also begun slashing payroll on a scale never before seen. In 2008, Citigroup cut seventy-three thousand people; Merrill Lynch and Bank of America dropped thirty-five thousand; all of Lehman's twenty-three thousand people were left jobless; nine thousand of Bear Stearns's employees weren't picked up by J.P. Morgan, which bought the investment bank for a pittance and eliminated another ten thousand jobs within its own house.

The difference for Conway was staggering. Where he once had to battle for each and every interview with a top-notch recruit, now he had a large pool of interested students. The same could be said for all other parts of the economy that once competed with Wall Street for bright quantitative minds.

"We started getting people that we would have never gotten before," recalls Glenn Kelman, the CEO of Redfin, a Seattle startup that is trying to reshape how we buy and sell real estate, undercutting normal broker commissions and the collusion that dominates the industry. Kelman has lamented out loud for years that so many of our country's smartest people were being locked up in lower Manhattan or Chicago's Loop to battle each other with clips of code flowing through fiber-optic cable. "It's insane that this is where most of them end up. Imagine what we could do if all that talent was freed," he says.

For Conway, the talent had indeed been let loose. In the two years following the banking implosion, eLoyalty hired sixty algorithm builders, ramping up his team to fit his lofty business aspirations.

During the decade that Conway has been building his personality-sensing bot, eLoyalty's traditional call center management work kept the ship running. During this period, the legacy business comprised 60 percent or more of the company's $90 million of annual revenues. It paid for Conway to construct what he views as an impossible barrier to wannabe competitors: a library of two million algorithms that listen, learn, and slot every person who speaks within earshot of the bot. Conway's bots have now analyzed more than 750 million conversations. Storing such data is a Google-worthy task. In 2011 eLoyalty finished off a new data center in Minneapolis that houses a thousand servers containing six hundred terabytes of customer data (the Library of Congress's print stash would require ten terabytes to store).

The bots' talents go beyond reading personalities and connecting us with like-minded people. They're fraud busters, and as they spread, they may help dampen what has become a global identity theft epidemic. One out of three thousand customer service calls is fraudulent. In these conversations, fraudsters try to milk customer service agents for personal information belonging to a target. In many cases the aspiring thief will only have one piece of a person's data, such as a name and address. They will make calls to hundreds of credit card, bank, and insurance vendors, hoping to gather more information as they go, and at some point hit upon an account they can manipulate for their own enrichment. These calls often get farmed out to call centers in Russia, Eastern Europe, Ireland, and Africa that are wholly dedicated to stealing information and identities. Fraudulent calls can often be spotted by a continuous string of wrong answers to security questions or, in most cases, the caller hanging up abruptly.

Operators will normally flag these calls as likely fraud attempts, but because the calls often carry disguised caller IDs, there's little that can be done to prevent them from placing hundreds or even thousands of additional calls. The eLoyalty bots' speech recognition capability gives them an uncanny ability to thwart fraud, however. People's voices are as uniquely distinguishable as a human fingerprint; they can't be easily changed, and their timbre and acoustics, while they change slightly over time, remain incredibly consistent. Conway's bots electronically record and store their callers' voices in a master database and can quickly check callers' voices in the fraudster voice database. Operators are instantly notified on their computer screens of a match and the high likelihood that a call is fraudulent. The crooks, as is their wont, are trying to pioneer voice-altering synthesizers to crack Conway's tool, but Conway is developing a counterhack, he says.

Beyond fraud, the bots are revolutionizing the annoying business of telesales. Different kinds of personalities require different selling techniques. Don't soft-talk a thoughts-based person. Feed these people the benefits, the facts, the savings, and hope the chips fall in the right spots. An actions-based person, however, may be well motivated by a pitch that emphasizes, "If you buy right now, I'll also throw in a second thing-amagadget for free!" The trick for the cold caller is knowing what kind of pitch to employ. Training telephone salespeople on the finer points of recognizing six different personalities isn't likely to be a profitable gamble. But being able to direct salespeople with the right pitch at the right moment is valuable.

Consider the experience of Vodafone, which also offered to be the guinea pig for sales pitches based on the methods developed at NASA and perfected by Conway's eLoyalty team. The telecom company's marketing department had a predetermined order and way in which their operators were to sell upgrades. Their methods, like most, were moderately successful, but most calls still resulted in strikeouts. But then Vodafone altered its pitches and selling order based on the personality of the potential customer. Emotions-based people would be buttered up with chummy talk and products that offered to keep them in better touch with relatives and friends. Other personality types had similar plans tailored to them. The result: an 8,600 percent increase in sales success.

The potential for bots in business, service, and even politics (think robocalls) is clearly huge. Conway believes it could be a multibillion-dollar business. Big companies are starting to see the promise of Conway's offering; several of the largest telecom companies in the country use his bot in some of their customer service seats. Four of the six largest health insurance companies have started using the bot as well. The rewards for Conway could be enormous. He charges $175 per seat per month for access to his algorithms, a pittance compared to the total cost of $50,000 per customer service seat for big companies. If a company like AT&T were to add Conway's bot to every one of its customer service chairs, the result would be $200 million in annual sales—more than double what his company pulls in right now.

If companies like AT&T do start switching over whole call centers to Conway's software, as many are now contemplating, many of our conversations in the future will be directed by algorithms. If Conway's plan is successful, we may find ourselves talking to eavesdropping bots on almost every business call. Why leave customers prone to the fickle treatment of their fellow humans when bots can secure more satisfaction, lower costs, and happier customers?

The effect of bots on our interactions with fellow humans won't end with phone conversations. Capers, the psychiatrist who introduced Conway to the methods and still works with them on a daily basis, believes that the science has bigger potential in places like schools and hospitals, where the efficacy of teacher-student and caregiver-patient relationships is acutely affected by the natural rapport or prickliness generated by the personalities on each side. Bots could perform quick two-minute assessments of patients and match them with nurses and doctors who fit similar molds, making for better communications and, ultimately, better care. The same thing can be said of the classroom, although those relationships can be harder to manage because of the size of schools, the number of students in a class, and so on. Capers believes that a large percentage of children diagnosed with attention deficit hyperactivity disorder are reactions-based people placed in settings (large schools) that don't provide the kind of learning environment—with more interaction, stimulation, and direct instruction—such personalities require.

Conway is looking at all of these avenues. His next target, after his bots conquer the world of voice communications, is e-mail. Businesses with large accounts often have client relationships more than thirty people deep on each side. All of these people trade e-mails, phone calls, and instant messages. To Conway, these exchanges and the words, thoughts, and messages they contain are intel. With it, he can derive the state of a relationship between two companies. He can identify tension points that may only be obvious to one side, business units that are endangering the client relationship, and even the odds that an account will be renewed. Employees' names and the actual words they've typed within e-mails remain anonymous to all but the bot that reads them, putting to use more than 100,000 algorithms to parse e-mails for things like tension, agreement, happiness, and conflict between two parties. Conway's algorithms provide a peek at the health of a relationship between two companies at any point, allowing executives to note, in some cases, what initiatives hurt or helped their goodwill with clients.

Allowing one's e-mails to be surveyed by an employer—even just a bot—doesn't sound like something most employees would happily accept. We shouldn't forget that one of the largest e-mail platforms in the world, Google's Gmail, does the same thing when it searches our e-mails for keywords and serves us corresponding ads. Not only that, but it's also worth noting that the law allows employers to do anything they wish—read, delete, forward, publish—with their employees' company e-mail accounts. In a world where an increasing amount of power is being accumulated by those who can fashion and leverage algorithms with skill and finesse, our future looks to be one filled with bots judging us, routing us, and measuring us.

Conway's technology, were it to take over, could theoretically limit the kind of people we talk to. Perhaps we'd only be allowed to interact with those who share our personality category. Such a fact translates to a better bottom line for companies using the software, but what would it mean for our culture and our capacity to accept others? It may not affect any of these things; exposure to people and cultures unlike our own can be had through any number of conduits. But couple the power of Conway's bots with the kinds of pigeonholing that technology has already brought to us, and the problem becomes more acute.

On the surface, such a condition may sound nice, but in reality it can lead to intolerance, extremism, and legislative stalemate. Technology has helped us build a world in which we needn't travel beyond our own comfort zone, where we can settle into an easy chair of like minds, souls, and views. If, as a thoughts-based person, an employee is only ever subjected to other brains of the same type, how will he handle a new coworker who is an emotions-based person? This employee may be one of the most talented people in the country at his job, but if he is unable to effectively communicate with people of differing personalities, where will that leave him as a professional? Working with other types of people is already one of the more challenging things we face in day-to-day life, but it can also be one of the most rewarding. Conway's mission hasn't sent him explicitly angling against such relationships, but as the bots take over different paths of human interaction, we must acknowledge the possibility of strange side effects on our world.

9

WALL STREET'S LOSS IS A GAIN FOR THE REST OF US

TO SILICON VALLEY THEY GO

IT TOOK A FINANCIAL CATASTROPHE to wrest the talent tide away from the bulls, bears, traders, and quants of Greenwich, lower Manhattan, London, and Chicago. As Conway built new teams with adept quantitative brains, so did the rest of the tech world.

The change in brain flow within the economy was noticeable to more than a few. "It was astonishing, what happened almost overnight," says Vivek Wadhwa, the Duke professor. His students' attitudes toward Wall Street began changing during the fall of 2008. Even better, from Wadhwa's point of view, their interest in technology and engineering picked up. "I believe that was the point of these degrees in the first place," he adds.

There's little doubt that the collapse of the financial industry changed the course of the economy. But just as with any great story, there are other branches of this tale. Wall Street's nadir of 2008 and early 2009 may have been the main act onstage, but there were other, more subtle plays taking place. One wasn't so subtle: Facebook. Mark Zuckerberg launched Facebook in early 2004, and he was at the time of this writing worth more than $25 billion. His company's value, meanwhile, has soared past $100 billion.

Zuckerberg's story is well known; the people who use Facebook aren't tech nerds or finance guys—they're exceedingly normal people, many of whom have become addicted to the constant updates, news, and chat that Facebook provides. Zuckerberg's story seems attainable; he's clearly smart, but he remains normal. I met him at an event to which he drove his girlfriend's Acura hatchback—a vehicle that was easily six years old. Compare that with the glitzy car stables belonging to much of the rest of the Silicon Valley elite, few of whom have collected a fortune even 1 percent of Zuckerberg's, and you're left with a story that affects college hackers and high schoolers alike. Who wouldn't want to be that guy, to build the next big thing from a dorm room and stay incredibly grounded throughout?

All Zuckerberg needed was the knowledge of how to code and how to construct an algorithm. A scene from the movie *The Social Network,* a profile of Facebook's rise, has been singed into the brains of millions of future algorithm writers, math majors, and programmers: Zuckerberg's character asks his buddy, Eduardo Saverin, for an algorithm Saverin used to rank chess players. Zuckerberg wanted to use the algorithm to rank girls at Harvard. Saverin scribbles the complex-looking equation onto a windowpane, some hipster music kicks in, and Zuckerberg nods: "Let's write it." The Harvard boys then set to work building a $100 billion company in six years. This moment, the scrawling of the algorithm on the window, may never have happened—*The Social Network* was nothing if not hyperbolic—but it's still an iconic scene and a myth of creation for many savvy young hackers.

Facebook deserves more credit than any other company for helping to swing the momentum of minds from Wall Street to the West Coast, but many other success stories in recent years have beaten anything even the greatest financial engineer can offer. Look at Twitter, Groupon, Square, Dropbox, Zynga, and YouTube, among a bevy of others. Then there are dozens of names, such as Tellme Networks,

Tacoda, Zimbra, and others, that you may never have heard of but that still fetched prices of $250 million to $900 million each when they were acquired. College students have noticed this trend, and the engineering and computer science schools of the country have seen their enrollment finally begin to tick up after years of decline. As Wadhwa noted, the best students are no longer fixated on Wall Street.

For quantitative minds with options, the tension within their career decisions has always been, do I take the guaranteed money on Wall Street—with starting salaries of $250,000 plus a bonus perhaps just as valuable—or do I move to an office with a kegerator and a ping-pong table while I swing for the fences of tech immortality? The scales are now tipping toward tech, even for experienced Wall Street hands.

Puneet Mehta was a senior vice president of technology at Citi Capital Markets, a position that paid about $400,000 a year, when in 2010, at age thirty-one, he chucked his Wall Street career for startup life. "I used to ask myself: am I really building something here?" Mehta says. Wall Street, he explains, exists to do one thing: get in between every single financial transaction it can. Before Citi, Mehta wrote code at Merrill Lynch and J.P. Morgan. The fact that on Wall Street all he would ever be was a middleman, albeit a well-paid middleman, used to keep him up at night. He began working on a side project, an app called MyCityWay, which allows urban dwellers to easily navigate restaurants, movie theaters, bars, traffic—anything that affects their life every day. When he reached his limit of writing code for Citi, Mehta quit and worked unpaid on his app. He and his cofounders now have fifty employees and their app has been released in seventy cities, thirty of them overseas. They've raised $6 million in venture capital, and Mehta says he talks to at least one other engineer every week who says they're never going back to Wall Street.

"It used to be hard to hire these people," he says. "Now they want to come."

Andrew Montalenti didn't need to be shoved out the door of Morgan Stanley. He just left in March of 2009 after spending three years as a hacker on Wall Street and thinking, "I never got the idea I was solving real problems."

Montalenti accepted a job at Morgan in 2006 halfway through his senior year at NYU, where he majored in computer science. The investment bank gave him a $10,000 signing bonus and a September start date, which Montalenti says made taking the job that much easier. All of the new recruits fall in love with the idea of spending the summer traveling through Europe or South America on Morgan's money, without a care in the world. It's not a bad pitch. When he finally started his job, Montalenti was making $100,000 at twenty-two years of age. At some point he looked around at his colleagues, some of whom were fifteen to twenty years older than him, and said, "This can't be my life's work." He quit and cofounded Parse.ly, whose algorithms put together nuanced and in-depth analytics for publishing Web sites like those for the *Atlantic* and *U.S. News and World Report,* both of which are clients. "We're helping editors and journalists," he says. "I can't imagine doing anything else."

Jeffrey Hammerbacher faced exactly the conundrum that Mehta, Montalenti, and thousands of other engineers confront when graduating. Hammerbacher was finishing up at Harvard with a math degree in 2005 when he made what seemed to be the easy choice to head to Wall Street.

Hammerbacher grew up in Michigan, the son of a General Motors assembly line worker. As a high school student, he composed poetry with a deft touch and snapped off curveballs with enough guile that the University of Michigan wanted him for its baseball squad. But he picked Harvard, where he started as an English major, later hopscotching, without a hitch, to math. College had difficulty entertaining him. At one point, Hammerbacher dropped out and headed home to work in a GM assembly plant before his mother's consternation propelled him back to Harvard.

Upon graduation, Hammerbacher's path to Wall Street was made even clearer by the fact that his girlfriend had already moved to New

York. "Everybody knew that Wall Street needed quantitative people—they recruited the math department at Harvard heavily," he explains.

Hammerbacher wasn't keen on becoming some quant analyzing derivatives and writing the next great trading algorithm. So when he joined his girlfriend in the city that summer, he had plans to start a math PhD at NYU in the fall. To kill time and make some money, he took an internship at Bear Stearns for the summer.

As happens so often on Wall Street, one little gold piece led to the promise of more, and Hammerbacher dropped his doctorate plans and signed on full-time with Bear. He found himself inside the churning heart of Bear's quantitative mortgage machine that worked tirelessly on algorithms and data crunching to support their myriad positions in the housing market. The housing boom had proven fantastically lucrative for all of Wall Street, and Bear, seizing that opportunity with gusto, had a small army of programming engineers and math savants cranking away on models that in most cases continued to back the now-we-realize-how-stupid-it-was idea of buying more mortgages from brokers, banks, and middlemen. As has been well chronicled in books such as Gregory Zuckerman's *The Greatest Trade Ever* and Michael Lewis's *The Big Short,* these mortgages that Bear bought were chopped up, mixed with Wall Street's version of gelatin, and repackaged into perfectly shaped wedges that could easily be sold to pension funds, foreign banks, and unsuspecting millionaires and billionaires.

Hammerbacher joined the quants working the mortgage desk at Bear and helped figure out, through stochastic calculus, regression models, and long, dynamic algorithms, how to properly hedge Bear's housing market risk. Hammerbacher tired of the game after a short nine months, however, and left in 2006. It was a prescient move, as Bear began showing weaknesses a year later as a pair of its hedge funds, heavily entrenched in the subprime mortgage market, began teetering toward insolvency. Bear would disappear altogether just twenty-four months after Hammerbacher quit, the firm's assets hocked to J.P. Morgan at rummage sale prices.

The gears of Bear's quant operations, whose math and programming work had been manipulated into endorsing what, we now can see, was an insane volume of mortgage-backed securities, had turned Hammerbacher off. If he wasn't going to pursue further learning within the structured setting of academia, he surely didn't want to waste his time on this.

Through a Harvard friend in the spring of 2006, Mark Zuckerberg reached out to Hammerbacher, who was well known for his math prowess. A week later, Hammerbacher was relocating to California. He was an early employee at Facebook—one of the first hundred—and Zuckerberg gave him the fancy title of research scientist. It was Hammerbacher's job to use math and algorithms to figure out how people were using Facebook, broken down by age, gender, location, and income, and why Facebook thrived in some places and flopped in others. Hammerbacher had been made king of Facebook's data.

Zuckerberg tapped Hammerbacher because the company was desperate to get a handle on the mounds of data that threatened to overtake its storage capacity. The nascent company knew the data was valuable, but it wasn't sure what to do with it. Wall Street has known what to do with such quantities of data for nearly two decades: store it, sort it, chop it, and look for patterns, anomalies, and trends that can be ridden to profit.

For all of Wall Street's data-sorting muscle, it's never faced a mountain like that which confronted Hammerbacher on the West Coast. Facebook's raw information piles make the grand datacenters of the NYSE and its high-frequency trading shops look quaint. For every one of the nearly one billion users in Facebook's system, the company stores up to a thousand pages of data, including the type of computer you use, your political views, your love relationships, your religion, last location, credit cards, job, favorite links, family, and, most voluminously, your photos.[1] That's to say nothing of our browsing habits or Facebook's vast advertisement infrastructure.

Facebook's first data center, in Prineville, Oregon, helps serve its users, who collectively spend more than ten billion minutes on the site each day while consuming one and a half million photos each second and generating thirty terabytes of data daily. Facebook also has two mammoth datacenters in North Carolina and is beginning construction on yet another, its first outside the United States, on the edge of the Arctic Circle in Lulea, Sweden. The frigid climate will help cool the thousands of computer processors and keep energy bills down. The three buildings, scheduled to come online in 2014, will cover an area equal to sixteen football fields. This is the world that Hammerbacher, who shortly before arriving was just another cog in the Wall Street quant machine, was brought in to tame.

Hammerbacher brought all that he learned on Wall Street to Facebook's Silicon Valley offices. He built a team stocked with thoroughbreds much like himself; some of them, having grown fed up on Wall Street, were plucked from other gold-plated firms like Morgan Stanley, Knight Trading, and Goldman Sachs. Facebook needed to know how to best arrange its pages, how to keep people connected, and, most important, how to keep people on its site as much as possible. "I was there to answer these high level questions, and they didn't have any tools to do that yet," he said of his hiring.[2]

Hammerbacher built the tools and the algorithms that still monitor the unimaginable amount of data pouring into Facebook every hour of every day. Part of the reason that Facebook has proven so "sticky" and irresistible to Web surfers is because Hammerbacher built systems to track people's mouse clicks, where their cursors stray, and what page arrangements hook the largest number of people for the longest amount of time. All of this click, eyeball, and cursor data gets strained, sifted, and examined. The stuff that proves stickiest stays; the rest goes into the digital dustbin.

Facebook says half of its users log in at least once a day. Nothing outside of e-mail platforms can make similar claims. The social networking site has proven so addictive that a meme circulating the Web features a clear syringe filled with blue liquid and adorned with the white Facebook logo. There may not be a better metaphor to describe the current surge of Internet companies and their brilliance in commanding so much time from so many people. Having a nearly captive audience of billions makes it all the easier, and lucrative, to sell ads that can be targeted by sex, income, geography, and more.

Facebook can tell prospective clients its users' tendencies, relationships, desires, and hobbies. Where most of the Web hands advertisers a blunt club, Facebook gives them a surgeon's scalpel. Minutely targeted advertising has become the engine of Facebook's business—and it became possible when a Wall Street quant defected from the dark side. Hammerbacher didn't migrate west for lack of money, status, or pride, but out of boredom and a general feeling that there had to be something more important he could be doing. He bought fully into the Facebook ethos of connecting the world. But even this work became, in his eyes, less and less important. "The best minds of my generations are thinking about how to make people click ads," he likes to say. "That sucks."

After two years at Facebook, Hammerbacher left. If we're to assume that Zuckerberg's company applies the standard rules of the Valley to all of its employee benefits packages—a safe assumption—then Hammerbacher would need four years at the company to earn his full vesting rights. That means that by the time he left, he'd only accumulated half of the shares Facebook allotted to him when he signed on as one of the company's early employees. So by quitting, Hammerbacher likely left behind tens of millions of dollars that were guaranteed to him. About that, he quipped, "An egregious act of wealth destruction."[3]

As the guy who organized the data of the largest user base on the planet, Hammerbacher is a known quantity as a hacker and math mind. Following his Facebook exit, he had no problem finding a position as an entrepreneur-in-residence at Accel Partners, an elite Silicon Valley venture capital firm and an early investor in Facebook. But Hammerbacher quit the VC scene after a month.

Having amassed some wealth and perspective at Facebook, Hammerbacher took a step back and surveyed the tech world. What could really, truly change the world? The answer, he realized, was a kind of general spade for digging through what the digital and computerized world produces in such copious quantities: data. Hammerbacher wanted to build a kind of operating system for sorting and swimming through labyrinthine sets of data produced by Web sites, utility companies, biology researchers, and health care providers. Shortly after having this epiphany, he founded Cloudera. Hammerbacher wants his platform to be as applicable for a cancer researcher as it is for a game company determining how to serve users with just the right amount of "buy" buttons. The Valley thinks he's on to something; Hammerbacher and the rest of the Cloudera founders have raised $76 million in venture capital, an outsized number for a software company so young.

Hammerbacher is one of a troupe that has shifted from lower Manhattan to Silicon Valley to help inject Wall Street–style data mashing into our everyday lives. Investment banks and trading houses learned to build their data mines into ramparts that can ward off competition, and now these kinds of defenses are being erected in Silicon Valley. "Facebook has more information about your clickstream than it does of photos of your friends," says Glenn Kelman, Redfin's CEO. "Web sites that have built scale are learning to dominate all others because they're the ones with the data to optimize."

Zynga, which makes the ubiquitous FarmVille game, was founded in 2007 by Mark Pincus. On the back of FarmVille, it has grown to be worth $20 billion, with more than 200 million users.[4]

Zynga's data stack is so thick and detailed that it has developed methods of how to best prod and poke different kinds of users in just the right way to get them to open their wallets. The company collects some sixty billion data points every day and knows which players are best hooked with ten minutes of free play before being prodded for a credit card, and which ones can be coerced into buying expensive virtual add-ons like fancy tractors, barns, and animal herds. The proof is in Zynga's sales and its data hoard. Kelman plans on doing the same thing for Redfin, which facilitates much of the house-buying process through its Web site (and agents that don't work on commission) rather than through real estate agents whose only motivation is a sale, no matter how good or bad the deal. Netflix has spent millions honing its algorithms that recommend what movies people should watch. Kelman points out that sorting the factors determining a person's preference of movie is easy when compared with sorting those affecting the purchase of a house. "If you watch a bad movie, you blew a Friday night," he says. "If you buy the wrong house, it can affect your life."

In early 2012, Kelman had just successfully recruited five Ivy League quant-hackers to join Redfin in Seattle. Two of them came from Bridgewater, the largest hedge fund in the world. Turning down Bridgewater is something that doesn't often happen; a few years there and you have a very real chance to be a millionaire. The only thing crazier, perhaps, would be to turn down a job at Renaissance Technologies. The Long Island operation is so full of high-level engineering and physics PhDs that admirers like to call it the "best physics department in the world." But at Y Combinator, the startup accelerator in Silicon Valley that continually draws in elite hacker talent, I met Ignacio Thayer, who, among other notable achievements, is the only person I've known to turn down Renaissance. Thayer had been a PhD candidate in computer science at Stanford when he interviewed with and was offered a job by the hedge fund. Before that, he had spent four years working on statistical translations for Google. Thayer eschewed hedge fund money and even finishing his PhD to start ReadyForZero, a Web site that helps consumers pay down their debts faster by employing bots that direct users, every month, on which debts should be paid first based on interest rates, balances, and the behavioral history of the user.

An employee of Renaissance, the highest concentration of elite quantitative talent in the financial world, is virtually guaranteed several million dollars, if not tens of millions, if they stay on for more than a few years. Employees—and nobody else—get access to Renaissance's Medallion Fund, which has famously racked up 30 percent or better gains since its inception in the early 1990s. The Medallion Fund employs algorithms trading millions of shares of, as founder Jim Simons puts it, "anything that moves." At its historical pace the Medallion Fund would turn $100,000 into more than $20 million in just twenty years. When exceedingly smart, calculating people begin turning down Renaissance riches for undetermined outcomes in Silicon Valley, the pendulum has officially swung 100 percent away from Wall Street.

"It used to be if you went to Harvard or Yale, you wanted to be a finance titan," Kelman says. "But now everybody wants to be a Zuckerberg."

THE DAMAGE WROUGHT BY WALL STREET

Where would our economy be without Silicon Valley and its software bots? As bad as the economy looked in 2008–2011, it would be far worse without the string of innovation that the Valley has continually dished out. We could likely even be better off had Wall Street not been sucking up so much of our country's technical talent—and the algorithms they built—for so long.

A 2011 study by the Kaufmann Foundation draws a relationship between a burgeoning financial sector during the 1980s, 1990s, and 2000s and a drop in the rate at which science and engineering graduates become entrepreneurs. The rate among experienced engineers, in fact, dropped by more than 60 percent since the early 1980s, when Wall Street started snatching up technical minds as fast as it could.[5]

The authors, Paul Kedrosky and Dane Stangler, write:

> The financial services industry used to consider it a point of pride to hire hungry and eager young high school and college graduates, planning to train them on the job in sales, trading, research, and investment banking. While that prac-

> tice continues, even if in smaller numbers, the difference now is that most of the industry's profits come from the creation, sales, and trading of complex products, like the collateralized debt obligations (CDOs) that played a central role in the recent financial crisis. These new products require significant financial engineering, often entailing the recruitment of master's- and doctoral-level new graduates of science, engineering, math, and physics programs. Their talents have made them well-suited to the design of these complex instruments, in return for which they often make starting salaries five times or more what their salaries would have been had they stayed in their own fields and pursued employment with more tangible societal benefits.

It's not often that the most important innovations in the world come from the GEs and the Microsofts, the authors point out. They come from entrepreneurs who are focused on that one area with an intensity that bigger companies simply can't bring. Most big firms see their time as best spent on making their current products and processes more efficient. Making a $100 billion company more efficient by just 5 percent could result in a $5 billion windfall to the bottom line; there are few new products that could ever create the same dent. Because of this fact, big companies tend to fall into traps of overmanaging and underinnovating. This is why startups are so important.

"Startup firms specialize—in a way that larger and more-established companies can barely contemplate—in attacking complex problems in cheaper and more efficient ways," write Kedrosky and Stangler. "For the leading areas in need of entrepreneurship today, scientists and engineers are essential to start firms or join new companies."

The good news is that people like Ignacio Thayer and Jeffrey Hammerbacher are part of a bigger trend that will continue, say the Kauffman report's authors: "Given our need for entrepreneurs to bring products and services to a market that help us with some of the most difficult and complex societal problems we have ever faced, there could not be a more auspicious time for the change."

Silicon Valley has always been ground zero for tech startups, followed by a basket of other cities. After long being an afterthought in the space, New York made a charge to the top of that second heap following the 2008 financial crash. It's an incredible turnabout—one that's no coincidence, says Wadhwa. "The 2008 banking dip has more than a little to do with that," he explains. "All of sudden you had tech talent on the streets—plus all of the graduating engineers who would normally move to New York and work for Wall Street. They still came, but they didn't go to work for the investment banks."

I asked Fred Wilson, a partner at Union Square Ventures and the most prominent venture capitalist in New York, if he thought the Wall Street collapse of 2008 directly kindled the now bustling startup scene in New York. Wilson thought there was more to it. "Startup ecosystems have long gestation periods," he says. "The seeds of what is happening now were sown in 1995 to 1999."

There's probably more than a little truth to both hypotheses. Wilson and the rest of the New York VC brood have a right to let people know they were pushing tech forward in New York when it wasn't hip to do so. Without them, we wouldn't be talking about this. But it clearly didn't hurt that thousands of highly motivated engineers with elite educations dropped into the mix thanks to Wall Street's misfortunes in 2008.

Wadhwa also credits the Occupy Wall Street movement of 2011 for adding a layer of grime to financial-sector prestige. Redfin's Kelman agrees, saying that many of the Ivy League science majors he used to recruit have simply been turned off to Wall Street.

"The rage that is brewing against investment banks is palpable in my undergraduate engineering classes," Wadhwa says. But he remains leery of Wall Street's capability to mount a comeback, pointing out that many of his students graduate with six-figure debts that need to get paid off somehow. To solve this conundrum, Wadhwa proposes a national endowment to pay for engineers' educations. Graduate from an

American university with an engineering degree and, presto, your debt gets paid off. But the professor would want to add a few caveats: If a graduate heads to Wall Street or into finance, the debt remains with the engineer. Those who head into true engineering fields or try their hand at a startup get their debts forgiven. "You'd be getting the cream of the crop in America to go after these scholarships," he reasons.

Paid or not, the programming skill of new graduates continues to climb. That only means more algorithms, more automation, and more changes to our lives.

10

THE FUTURE BELONGS TO THE ALGORITHMS AND THEIR CREATORS

WITHIN GROUPS OF FRIENDS, THERE are always pecking orders. At the top sit the influential ones who shape the opinions of others, the leaders. At the back sit the followers. The same is true at work or within a family. There are bosses and there are plebes. Everybody knows this on some level and most people have a sense of where they think they fit. But that sense is corruptible. You may feel like a leader, but in fact you're a lackey. Your friends know it, you don't. Some people float to the top based on their position, their money, their looks, or just sheer popularity. These are the true wielders of influence in society. Who these people are can sometimes be obvious. But often it isn't so apparent.

There exist ways to find such things out. How we speak indicates what spot we occupy in the pecking order. The influencers rarely change their speech patterns. Those being influenced, however, often alter their speech to be closer to that of the influencer. The same applies to written communication such as e-mail. Words betray one's influence or lack thereof. He who dictates style is at the head of the pecking order.

He who shapes his language more like the other is lower. More specifically, our hierarchy can be determined by how we use words that are articles, auxiliary verbs, conjunctions, and high-frequency adverbs.[1] These crumbs **of** grammatical fodder are **what make up** our style, **be it in the** way we speak **or in the** way we talk. Such words are bolded in the previous sentence.

Jon Kleinberg, a computer science professor at Cornell, wrote an algorithm that keys on our word use to determine who among us are real influencers, the people who shape opinions, drive trends, and command the attention of others. These people can decide elections or propel a new product. The algorithm listens in on conversations and identifies who is in charge and how the rest of the pecking order plays out. It also works well on e-mail strings. Imagine what would happen if this algorithm eavesdropped on all communications going through Google or Yahoo! or the e-mail servers of GE or Goldman Sachs. The influencers could immediately be identified. A company that knew exactly who influencers were could tailor their marketing toward them. On the Internet, influencers would be followed by banner and text ads for products that were, unbeknownst to them, priced much higher than average ads. Companies could use what are called browser cookies to track, advertise, and keep tabs on the influencers. These people are the gateways to others' hearts and minds, and the marketing world knows it. Being able to identify pecking-order top dogs automatically on a wide scale could lead to new approaches in politics, management, sales, and marketing.

This model of ranking things based on small clues of influence is the same calculus that drives PageRank, Google's algorithm, named after cofounder Larry Page, which steers Web traffic to sites the Web regards as authoritative on the subject being searched. Important Web sites are called hubs and influencers. Google gives more credence in its search results to sites that are often linked to by influential sites and hubs. If these sites commonly refer to, say, a particular flight-booking search engine as the best one while concurrently linking to it, it's likely that this Web site will rise to the top of Google's results. By looking at where the influential sites link, Google's algorithm can quickly determine what to show for any query a user might type in. Sorting humans can be done much the same way. Long before he tackled the subject of human influencers, Kleinberg wrote an innovative Web search algorithm while at IBM in the 1990s. It trawled the Internet looking for what he called hubs and authorities—later inspiring the Google founders to do the same thing.

Putting this algorithm into everyday use could prove a bit scary. What if Facebook dropped it on its entire site, ranking people according to their relative power within their circle of several hundred friends? What if companies did this, choosing managers by who already controls whom rather than whatever meritocratic process they've tried to install? Some things are just better left unsaid, be it by an algorithm or a human. But such quantitative measurements of our places in society are undoubtedly coming. As soon as Kleinberg created this algorithm, there were likely ten companies setting out to make their own. There's too much money to be made by marketing agencies, advertisers, and anybody interested in wooing the most important people in society for this not to happen. We face a future where rigorous and cold calculations such as this will constantly take place at the hands of algorithms. Pecking orders are just the beginning.

ALL ROADS . . .

If you're keeping track, algorithms already have control of your money market funds, your stocks, and your retirement accounts. They'll soon decide who you talk to on phone calls; they will control the music that reaches your radio; they will decide your chances of getting lifesaving organ transplants; and for millions of people, algorithms will make perhaps the largest decision in their life: choosing a spouse. At least they don't drive.

Or do they? After I relocated to California for the summer of 2011, I often found myself driving along Silicon Valley's Highway 85 and Interstate 280. At least once a week, I would pull alongside strange Toyota Priuses affixed with large rotating doodads on their roofs. There was always at least one human in the car, but he (and it was always a he) didn't have his hands on the wheel and wouldn't respond to our catcalls or questions. These cars, it turns out, belong to Google. People don't drive them, algorithms do. That swiveling thing on the roof is a laser rangefinder. The Priuses, seven of which were plying the roads in 2011, also sport GPS receivers, four radars, a video camera, a position estimator, and an internal motion sensor—all of which are wired into a master bot that makes decisions about how fast to go, when to turn, and when to stop. All of these systems together—the multiple radars, cameras, and GPS—provide redundancy. If the bot is getting data from one of them that wildly contradicts that being supplied by other sources, the variant data will be ignored, keeping the car from suddenly stopping or swerving because one radar was fooled by a swooping flock of birds.

Google's cars, which drive utterly autonomously from their human passengers, have covered 140,000 miles so far without an accident. A widely reported crash involving a Google car in August 2011 occurred when the human inside the car had been at the controls. Google's stated mission is to reduce emissions (bots are smoother on the gas pedal), double road capacity (bots need less time to react to trouble than do humans, so they can drive closer together), and cut in half the world's 1.2 million traffic deaths per year. Human error, which leads to almost all of the thirty-three thousand deaths on U.S. roads every year, could be almost eliminated if algorithms were the drivers.

As our world shifts from one where humans have made all of the important decisions to one in which we share that role with algorithms, the value of superior intellect has increased at a compounding rate. Software and computer code can now easily amplify the power of a small group of dynamic minds, increasing their potential value to society. These are the minds that are changing life on earth, taking critical decisions, from driving cars to trading stocks to choosing what personalities belong in a company, out of the hands of humans and putting them into the hands of bots.

A TALE OF TWO COASTS

Jeffrey Hammerbacher is one of the linchpins in the tale of quant talent drifting away from the steady riches of Wall Street for the unpredictable and meritocratic zone of Silicon Valley. He's a symbol of all that's different between the two worlds. One side uses bots and algorithms to chop up subprime mortgages and peddle them to unsuspecting German banks that end up eating the debt for pennies on the dollar—sorry about that!—and to trade stocks at fast speeds that do in fact benefit the public in lower transaction costs, but also cost them in the way of higher volatilities and the possibility of epic chaos. The other side, Silicon Valley, uses similar brains and technology to create a game that may keep you entertained for fifteen minutes a day. The best products from the Silicon Valley help you keep in touch with relatives, trace your family's origins, teach your daughter calculus, or smooth your relationship with your health insurer. That's what Hammerbacher saw and why he founded Cloudera.

When you ask Hammerbacher what he sees as the most promising field that could be hacked by people like himself, he responds with two words: "Medical diagnostics." And clearly doctors should be watching their backs, but they should be extra vigilant knowing that the smartest guys of our generation—people like Hammerbacher—are gunning for them. The targets on their backs will only grow larger as their complication rates, their test results, and their practices are scrutinized by the unyielding eye of algorithms built by bright engineers. Doctors aren't going away, but those who want to ensure their employment in the future should find ways to be exceptional. Bots can handle the grunt work, the work that falls to our average practitioners.

To the delight of many and the chagrin of a few, lawyers are as susceptible as anybody to an algorithm invasion. One of the most expensive parts of any large lawsuit is the process of discovery—collecting and studying all relevant documents in a case. In corporate law or contentious cases involving contracts, there can be millions of pages that need to be read. The readers are often paralegals or junior associates, either of whom will cost north of $100 or sometimes $200 an hour for, essentially, the service of their eyes. Eyes, though, can be hardware, and the brain behind the eyes can be an algorithm. A Palo Alto company called Blackstone Discovery, among others, is replacing people with algorithms that work just as efficiently, and often more thoroughly, than humans. Algorithms are cheaper too. Running through 1.2 million documents, a task that could once have cost $5 million or more, now costs only $100,000.[2] Perhaps that helps explain why in 2012 there were only twenty-six thousand jobs waiting for the fifty-four thousand or so lawyers who pass bar exams in the United States. None of this is good news for the tens of thousands of lawyers in the country who now want for work. *Oh, sorry, we gave that job to an algorithm.* The algorithm does not require health insurance or the promise that it could make partner later.

Doctors, lawyers, psychiatrists, truck drivers, musicians—just how many jobs might we lose to algorithms? How much will this affect our economy? In a landmark paper, two economists at MIT, Eric Brynjolfsson and Andrew P. McAfee, wrote, "Many workers, in short, are losing the race against the machine."[3] The median worker, the average white-collar accountant, the MIT pair warn, should prepare to be replaced.

Academics have made such predictions before. When machines began taking over manufacturing tasks in the 1920s and 1930s, John Maynard Keynes sounded the alarm for a "new disease" he termed "technological unemployment," which happens when jobs can't be replaced as fast as they're eliminated by automation.[4] Keynes's warning was blown off as hyperbolic when it didn't prove out. But perhaps his theory was simply ninety years early. Since the end of the recession in

June 2009, according to Brynjolfsson and McAfee, corporations have spent 26 percent more on technology and software but haven't raised their payrolls at all. In 2011, the S&P 500 logged record profits of nearly $1 trillion, a feat that didn't take many extra people to pull off.

Writers aren't safe either. Narrative Science, a company in my hometown of Evanston, Illinois, has attracted $6 million of venture capital for its bots that transform raw stats into styled, grammatically correct, and clever news stories. The sports Web site belonging to the Big Ten Network uses the technology to post articles within a minute after a game is over. Its algorithms suck in the box score and identify the most important parts of the game and then write a story built around those moments, just as a journalist would. The algorithm even knows what piece of the game is the best candidate to use as the "lede." An article duly punched out by the bot seconds after the third quarter of a Wisconsin-UNLV football game started like this: "Wisconsin appears to be in the driver's seat en route to a win, as it leads 51–10 after the third quarter. Wisconsin added to its lead when Russell Wilson found Jacob Pedersen for an eight-yard touchdown to make the score 44–3."[5]

So what's the lesson? Get friendly with bots. The people who are and will remain most indispensable in this economy are those who can build, maintain, and improve upon code and algorithms. Spend any time in Silicon Valley and the one thing you will hear technology companies large and small complain about is the dearth of good engineers.

Engineers with programming skills do not want for work. It's for this reason more than any other that companies outsource engineering work to places like India, Brazil, and Eastern Europe. If there were more qualified science, math, and engineering majors, the economy would have jobs for them, and those jobs are not likely going away. Even in the face of that fact, every year we continue to lose 40 percent of our prospective engineering majors to other, usually less math-intensive degrees. If this sounds like a familiar theme, it is. In 2011, President Obama called for the country to graduate ten thousand more engineers a year, an admirable goal.[6] And we've made some progress. Getting more kids interested in math and science at the high school level isn't the problem; there's been a successful push in many schools across the country to do just that.

The problem is often that kids get to college and run into what David Goldberg, an emeritus engineering professor at the University of Illinois at Urbana-Champaign, calls "the math-science death march," usually consisting of three semesters of relentless chemistry, physics, math, computer science, and, at Dr. Goldberg's school, a series of classes called theoretical and applied mechanics (TAM). Having run Dr. Goldberg's gauntlet myself, I can attest to sleepless nights and difficult tests, at least compared to the more transferable interdisciplinary classes and majors some humanities students take. Having said that, getting through four years of engineering school doesn't take genius; it just takes a little more work and a good base of upper-level math and science in high school.

The problem, unfortunately, is that not enough U.S. kids get that foundation of upper-level math before arriving at college. Some, including Renaissance Technology's founder Jim Simons, say part of the solution would be to hire more of the types of minds that are responsible for our coming algorithm-centric future and put them in the classroom. People can only teach what they understand themselves. Teaching isn't easy, however. Many of the skills a Google engineer holds do not transfer to unruly collections of sixteen-year-olds. We should seek out those special people who can comfortably go back and forth between these two worlds—from piles of algorithmic code to the high schools where the next generation of minds is being molded. Finding technical talent with the nuanced ability to rally students around new, complex subjects is a rare thing. But every time such talent can be spotted and placed into the educational field, even if it costs more money, it's a win for society and for students.

So does this mean creating yet another bidder—schools—for our best quantitative minds? It does. And that's okay—there are a lot of potentially quantitatively minded people roaming around out there who have never given their brains a proper crack at the game. Smart people aren't in short supply. Smart people educated in quantitative fields are, however. We just need to increase the size of the funnel that gets people there. Every single student at every high school in America should be required to take at least one programming class. Most students will stop there and move on to do something else. But even if just 5 percent of those students become engaged with the power of devising their own programs and algorithms, it will change the dynamic of our education system and our economy. Imagine all of the students who never give programming or quantitative fields a thought. Math, to them, is a rote skill that must be memorized so that a test or a quiz can be passed; they never see the other side of math that's changing our world. Or when they finally do, perhaps in college, their life vector is already set toward another field. Programming and computer science classes shouldn't be relegated to a niche group of students—this is the skill, more than any other, that will matter during this century. All students should get their chance.

There will be two large growth engines in the economy of the next fifty years: health care and tech. The former is the result of a system built on a shaky foundation and an aging U.S. population. But tech offers the opportunity for anybody from any background to attain skills that will keep them employed in good economies and bad ones. For a select few willing to ignore risks and carve their own path, technical skills and the ability to devise innovative algorithms can open the door to entrepreneurship and creativity. The door is open for anybody who wants to walk through it. Being a technical whiz isn't about scoring well on the math and science sections of standardized tests. It's about practice. It's about putting in time to learn processes.

There's going to be a lot of work in the future for those who can write code. If you can also conceive and compose intricate algorithms, all the better—you may just be able to take over the world. That is, if a bot doesn't do it first.

ACKNOWLEDGMENTS

It's remarkable how much can change from when a book project starts to when it ends. Sources, advisers, and even editors shift into and then out of the picture as the months roll into, in this case, two years. The writer remains the same, of course—as did my agent, David Fugate of Launch Books, who, as always, remains my number one advocate, champion, and adviser. Meeting David randomly on the Web—more than a year before I began work on my first book—changed my career. Despite the fact that this is my second book, I remain something of a noob in the publishing business. And David, quite simply, is the reason I can safely carry on as a noob.

A number of people pored over these pages before anybody who may have paid money for this book ever saw them. At the head of this group was Niki Papadopoulos, my editor at Penguin. Niki did not start as this book's editor, so she had little idea what she was wading into. But I got lucky. Niki is a superb editor, and she quickly helped shape this book's narrative, shave fat, and suss out weaknesses. The book is far stronger as a result of her work.

My wife, Sarah, read every word in this book—oftentimes more than once—and rid these pages of embarrassing clichés, grammar faux pas, and meaningless clauses. Life's forces often seem to collude in pushing many personal, family, and work events onto the same calendar. The last year for us has proved this true. Sarah gave birth to twin girls (hi, Parker and Maren!) just as this book was getting wrapped up, so her work was often done with one baby in her lap, another at her feet, and our son (hi, Jack!) swinging from a piece of furniture in the background. By the time this book is in print, Sarah will be back at her job and duly enshrined into the Supermom Hall of Fame. Without her, I would have crumbled many times during this project. She is my partner in work and life.

Marcel Pacatte, one of my professors from Northwestern University who was worth the price of tuition, continues to be a tireless source of critiques and advice. He read every chapter before anybody else and offered honest commentary that improved the manuscript at each turn.

Tom Pennington, my trusty friend in engineering school, skiing, pizza making, and life, dropped everything on a weekend to read through the pages of this book just because I asked him to. His feedback was invaluable.

My parents, Gary and Janet, also read sections of this book before it reached others, and I will always be encouraged by their enthusiasm for my work. My father is one of the few people who may have known most of chapter 2's content on the history of algorithms before reading it here. Nonetheless, he gave it a hearty stamp of approval, which quieted my worries that the chapter would bore people. He also designed my house, for which I can never thank him enough, so I do so here.

I didn't know Randy Stross, a technology columnist at the *New York Times* and a professor at San Jose State University, before May 2011. But he took to reading an early version of this book with gusto. His feedback was frank and profuse—and incredibly helpful. When Randy ran across a set of words he felt to be egregiously wrong for any number of reasons, he would lightly embed this word in the text: **"UGH."** Randy's "UGHs" improved this book and helped me continue my jagged evolution as a writer.

Thomas Peterffy was incredibly generous in spending more than a dozen hours with me in Greenwich, Connecticut, as well as on the phone. He was always a good sport when it came to my queries, and his colorful stories helped bring to life the first big hack of the U.S. stock markets.

Daniel Spivey and David Barksdale, the founders of Spread Networks, have given very few interviews to anybody. I was lucky enough to break their story nationally, and when, a week later, a competing reporter at the *Wall Street Journal* called them, they had nothing to say to the newspaper. Their story and their project remain almost unbelievable—and fun. I'm grateful that they, along with David's father, Jim Barksdale, a technology legend by any measure, took the time to tell me how they dug their secret hole.

Ben Novak has one of the most interesting stories in this book, and he took time from New Zealand to hop on a few Skype calls and tell me about it. Mike McCreary, whose technology made Novak's music popular, was generous in providing the history of his company as well as his own history, which is dotted with inventive creations.

David Cope, the creator of algorithms that can mimic the composition style of Bach, took hours to speak with me about his work. For anybody who is looking into the world of computer science as an outsider and wondering what it would take to get inside, Cope's tale is instructive. He was a musician and academic who, with zero experience, turned himself into one of the preeminent creators within the artificial intelligence community. His stories could constitute a book on their own.

Jason Brown talked to me about how he cracked Beatles mysteries that were four decades old. He explained the construction of chords, the science of sound, and the math behind music to a nonmusician. He was patient, and I'm thankful for that.

Bruce Bueno de Mesquita has a wonderful laugh and a knack for explaining how his game theory systems can be more effective than humans when determining what, exactly, other humans are most likely to do in the future. After I interviewed him, he appeared on the *The Colbert Report* in one of that show's more hilarious segments. I recommend watching that or, just as good, Bueno de Mesquita's TED lecture or his appearance on *The Daily Show*. A quick Google search will reveal all of these items.

Carnegie Mellon professor Tuomas Sandholm has built a reputation as one of the most advanced creators of game theory algorithms in the world. He built a company on these theories to employ more than one hundred people, then sold it, and has now found a better way to build poker-playing bots and, more important, to match organ donors. His work saves lives. He maintains a busy schedule, but he took time to talk with me at length on multiple occasions.

Dr. Terry McGuire is one of the two or three most interesting people I have ever spoken with. He was a stabilizing force at NASA for decades while serving as the agency's main psychologist. He has been chased by Soviet commandos behind enemy lines and was a confidant to men like John Glenn. Before there was the equipment that we know today for rock climbing and mountaineering, McGuire scaled some of the thorniest peaks in North America with world-class climbers. His insights on how our minds work and the way in which words betray our personalities and true intentions form the basis of a chapter of my book and a new class of algorithms that is invading all mediums of human communication.

Hedges Capers says the same about Terry McGuire, and his insights on the topic proved to be just as important for me. I marveled at how quickly he analyzed my own personality from the words I spoke. His analysis of my own tendencies as a person was quite accurate.

Kelly Conway spent hours with me on more than one occasion to detail how his company—then called eLoyalty and now Mattersight Corporation—constructed algorithms to, as he says, categorize the human language. Conway's company has now focused on this mission alone, and, by the looks of the value of his company's shares on the Nasdaq, his plan is working out incredibly well after some rough waters early on. Conway is an innovator and an excellent communicator in his own right, and I'm indebted to him for the time he spent with me.

Vivek Wadhwa, a professor at Duke University, was always up for a chat on Wall Street versus Silicon Valley—his views, gleaned from both sides of the talent war, helped shape that chapter. Jeff Hammerbacher isn't the only engineer to defect from New York to the Bay Area, but he's one of the most important. As the founder of a hot startup, he has few spare minutes in his life, and I thank him for giving some to me. The same can be said for Redfin CEO Glenn Kelman, whom I met years ago and whose pithy advice at that time helped convince me to write my first book. His thoughts were instrumental in painting the right picture of the Silicon Valley scene right now.

Many of the books I needed for researching this work, especially the ones on algorithmic trading and finance, were quite expensive. As an author, I like to buy books whenever I can, but many of these volumes, put together for aspiring Wall Street algorithm writers, cost a hundred dollars or more. Luckily for me, I live in the same town as Northwestern University, where the huge library houses every single arcane book I've ever sought. Also luckily for me, Northwestern gives alumni full access to its stacks and resources, affording me a vast and quiet workshop that was open early, late, and at all hours in between. When putting together the chapter on the history of algorithms, I found myself seeking old texts, often out of print and hard to find. Again, Northwestern had all of these materials, and I was able to browse them at will. The librarians there even waived some hefty late fees I had accrued on several books. Thank you.

I started this book and wrote much of it as a staff writer at *Forbes* magazine. By the time I finished, however, I had left *Forbes* to form a startup, Aisle50, which offers grocery discounts to consumers. It was quite a change for me but also one that I embraced. There have been many helping hands along the way, some of the most formidable ones coming from our investors and advisers at Y Combinator, Paul Graham and Jessica Livingston. They have built something special in Silicon Valley, and, for the curious, there happens to be a book being released at exactly the same time as this one, by the same publisher, that is the best chronicle ever put together on Y Combinator: Randy Stross's *The Launch Pad.* Read it.

As for Aisle50, I have high hopes thanks to our crack sales and engineering teams, who have proven to be up to every challenge we faced so far, and there have been many. Most important, I'm grateful to Riley

Scott, my cofounder. Other than the relationship you share with your spouse, there is none as intense as the relationship you share with your cofounder. We've lived together, worked together fifteen hours a day for months on end, laughed together, and even cried together. Riley is a tireless worker, a great husband, father, and friend—and there's nobody with whom I'd rather be in business. He's a strong person whose life has not been short on challenges, some of which he willingly courted as an ambitious person, and some of which, including the most harrowing, he did not. He has met the turns of life with a strength that few could have mustered. It's inspiring.

NOTES

CHAPTER 1: WALL STREET, THE FIRST DOMINO

1. Carolyn Cui, "From Healing to Making a Killing," *Wall Street Journal Asia,* April 27, 2010, p. A9.
2. Suzanne McGee, "A Breed Apart," *Institutional Investor,* November 10, 2005.
3. Joe Klein, "Sweet Sweetback's Wall Street Song," *New York,* September 5, 1983, p. 43.
4. Ibid.
5. Peterffy still keeps one of the tablets in a wooden display case in his office.
6. Thomas Bass, *The Predictors* (New York: Henry Holt, 1999), p. 126.
7. Felix Salmon and Jon Stokes, "Algorithms Take Control of Wall Street," *Wired,* December 27, 2010.
8. Hal Weitzman and Gregory Meyer, "Infinium Fined $850,000 for Computer Malfunctions," *Financial Times,* November 25, 2011.
9. Leo King, "Rushed Software Testing Results in Unprecedented Fine for Futures Giant Infinium," ComputerWorldUK.com, http://www.computerworlduk.com/news/it-business/3322223/rushed-software-testing-results-in-unprecedented-fine-for-futures-giant-infinium/.

CHAPTER 2: A BRIEF HISTORY OF MAN AND ALGORITHMS

1. http://news.ycombinator.com.
2. Jean-Luc Chabert, ed., *A History of Algorithms: From the Pebble to the Microchip,* translated by Chris Weeks (Berlin: Springer-Verlag, 1999).
3. The author was assigned the tic-tac-toe machine problem himself as an undergraduate engineer in a C-language programming class.
4. Chabert, *A History of Algorithms.* This introductory section relies heavily on the work of Mr. Chabert.
5. Ibid.
6. Godfried Toussaint, *The Euclidean Algorithm Generates Traditional Musical Rhythms* (Montreal: School of Computer Science, McGill University, 2005), http://cgm.cs.mcgill.ca/~godfried/publications/banff.pdf.
7. Midhat J. Gazale, *Gnomon: From Pharaohs to Fractals* (Princeton, NJ: Princeton University Press, 1999), p. 33.
8. Niall Ferguson, *The Ascent of Money* (New York: Penguin, 2008), p. 34.
9. Henry Linger, ed., *Constructing the Infrastructure for the Knowledge Economy,* Proceedings of the 12th International Conference on Information Systems and

Development, Melbourne, Australia, 2003 (New York: Kluwer Academic/Plenum Publishers, 2004).

10. "Apple and the Golden Ratio," *Paul Martin's Blog*, http://paulmmartinblog.wordpress.com/2011/07/18/apple-and-the-golden-ratio/.
11. Ferguson, *The Ascent of Money*, p. 34.
12. Dirk Struik, *A Concise History of Mathematics* (Mineola, NY: Dover, 1948), p. 80.
13. David Berlinski, *The Advent of the Algorithm: The 300-Year Journey from an Idea to the Computer* (New York: Mariner, 2001), p. 14.
14. Ibid., p. 2.
15. Bertrand Russell, *A Critical Exposition of the Philosophy of Leibniz* (New York: Cosimo Books, 2008), p. 192.
16. Chabert, *A History of Algorithms*, p. 44.
17. Bruce Collier and James MacLachlan, *Charles Babbage and the Engines of Perfection* (New York: Oxford University Press, 1998), p. 46.
18. Berlinski, *The Advent of the Algorithm*, p. 3.
19. William Ewald, *From Kant to Hilbert: A Source Book in the Foundations of Mathematics* (Oxford: Oxford University Press, 1996), p. 446.
20. Paul Nahin, *The Science of Radio* (New York: Springer-Verlag, 2001), p. xxxvi.
21. David Berlinski, *Infinite Ascent: A Short History of Mathematics* (New York: Modern Library, 2005), p. 45.
22. Nicholas Jolley, ed., *The Cambridge Companion to Leibniz* (Cambridge: Cambridge University Press, 1995), p. 251.
23. Richard Lindsey and Barry Schachter, *How I Became a Quant: Insights from 25 of Wall Street's Elite* (Hoboken, NJ: John Wiley & Sons, 2007), p. 126.
24. Stephen M. Stigler, "Gauss and the Invention of Least Squares," *Annals of Statistics* 9, no. 3 (1981): 465–74.
25. Jyotiprasad Medhi, *Statistical Methods: An Introductory Text* (New Delhi: New Age International Publishers, 1992), p. 199.
26. Jagdish K. Patel and Campbell B. Read, *Handbook of the Normal Distribution* (New York: CRC Press, 1996), p. 4.
27. Michael Bradley, *The Foundations of Mathematics* (New York: Chelsea House, 2006), p. 5.
28. Ioan James, *Remarkable Mathematicians: From Euler to von Neumann* (Cambridge: Cambridge University Press, 2002), p. 58.
29. Jane Muir, *Of Men and Numbers: The Story of the Great Mathematicians* (New York: Dodd, Mead & Company, 1961), p. 158.
30. Ibid., p. 159.
31. Elena Prestini, *Applied Harmonic Analysis: Models of the Real World* (New York: Spinger-Verlag, 2004), p. 99.
32. Michael Bradley, *The Foundations of Mathematics* (New York: Chelsea House, 2006), p. 20.
33. Bernhard Fleischmann, *Operations Research Proceedings 2008* (Berlin: Springer-Verlag, 2009), p. 235.
34. Keith Devlin, *The Unfinished Game: Pascal, Fermat, and the Seventeenth-Century Letter That Made the World Modern* (New York: Basic Books, 2008), p. 5.
35. Michael Otte, *Analysis and Synthesis in Mathematics: History and Philosophy* (Dordrecht, Netherlands: Kluwer Academic Publishers, 1997), p. 79.

36. Stephen M. Stigler, *The History of Statistics: The Measurement of Uncertainty before 1900* (Cambridge, MA: Harvard University Press, 1986), p. 5.
37. Romain Rolland, *Handel* (New York: Henry Holt, 1916), p. 108.
38. Robert Bradley, *Leonhard Euler: Life, Work, and Legacy* (Amsterdam: Elsevier, 2007), p. 448.
39. William Dunham, *Euler: The Master of Us All* (Albuquerque, NM: Integre Technical Publishing, 1999), p. xx.
40. Charles Gillespie, *Dictionary of Scientific Biography* (New York: Charles Scribner's Sons, 1976), p. 468.
41. Robert Bradley, *Leonhard Euler,* p. 412.
42. David Richeson, *Euler's Gem: The Polyhedron Formula and the Birth of Topology* (Princeton, NJ: Princeton University Press, 2008), p. 86.
43. Howard Rheingold, *Tools for Thought: The History and Future of Mind-Expanding Technology* (New York: Simon & Schuster, 1986), p. 39.
44. Ivor Grattan-Guinness and Gérard Bornet, eds., *George Boole: Selected Manuscripts on Logic and Its Philosophy* (Basel: Birkhäuser Verlag, 1997), p. xiv.
45. Margaret A. Boden, *Mind as Machine: A History of Cognitive Science* (New York: Oxford University Press, 2006), vol. 2, p. 151.
46. Anne B. Keating and Joseph Hargitai, *The Wired Professor: A Guide to Incorporating the World Wide Web in College Instruction* (New York: NYU Press, 1999), p. 30.
47. Ibid., p. 38.
48. Ibid.

CHAPTER 3: THE BOT TOP 40

1. Malcolm Gladwell, "Annals of Entertainment: What If You Built a Machine to Predict Hit Movies?" *New Yorker,* October 16, 2006.
2. Ibid.
3. Claire Cain Miller, "How Pandora Slipped Past the Junkyard," *New York Times,* March 7, 2010.
4. *Sunday* magazine documentary show, TV New Zealand, August 2008, http://tvnz.co.nz/sunday-news/sunday-2338888. Also see YouTube: http://www.youtube.com/watch?v=ilFEt2wpYck.
5. Jeff Chu, "Top of the Pops," *Time,* March 19, 2001.
6. Ryan Blitstein, "Triumph of the Cyborg Composer," *Miller-McCune,* February 22, 2010, http://www.miller-mccune.com/culture/triumph-of-the-cyborg-composer-8507.
7. Ibid.
8. Gerhard Nierhaus, *Algorithmic Composition: Paradigms of Automated Music Generation* (New York: SprinerWienNewYork, 2009), p. 1.
9. David Cope, *Virtual Music* (Cambridge, MA: MIT Press, 2001).
10. Ibid.
11. Chris Wilson, "I'll Be Bach," *Slate,* May 19, 2010.
12. Blitstein, "Triumph of the Cyborg Composer."
13. George Johnson, "Undiscovered Bach? No, a Computer Wrote It," *New York Times,* November 11, 1997.

14. Ibid.
15. Ibid.
16. Ibid.
17. D. H. Cope, *Comes the Fiery Night: 2,000 Haiku by Man and Machine* (Santa Cruz, CA: CreateSpace, 2011).
18. Walter Everett, *The Beatles as Musicians: The Quarry Men through* Rubber Soul (Oxford: Oxford University Press, 2001), p. 77.
19. Steven D. Stark, *Meet the Beatles* (New York: HarperCollins, 2006).
20. "George Harrison Webchat," http://www.georgeharrison.com/#/features/george-harrison-webchat.
21. Geoffrey Poitras, *Risk Management, Speculation and Derivative Securities* (New York: Academic Press, 2002), p. 454.
22. Rajendra Bhatia, *Fourier Series* (New York: Mathematical Association of America, 2005), p. 11.
23. Jason Brown, "Mathematics, Physics and 'A Hard Day's Night,'" *Canadian Mathematical Society Notes,* October 2004.
24. J. S. Rigden, *Physics and the Sound of Music* (New York: John Wiley & Sons, 1977), p. 71.
25. Brown assumed that an A note carried a frequency of 220 Hz. His conversion function looked like this: $f(x) = 12 \log_2(x/220)$.
26. George Martin, *All You Need Is Ears* (New York: St. Martin's Press, 1979), p. 77.
27. Jason Brown, "Unraveling a Well Woven Solo," http://www.jasonibrown.com/pdfs/AHDNSoloJIB.pdf.

CHAPTER 4: THE SECRET HIGHWAYS OF BOTS

1. Roy Freedman, *Introduction to Financial Technology* (New York: Academic Press, 2006).
2. Richard Bookstaber, *A Demon of Our Own Design* (New York: John Wiley & Sons, 2011).
3. Freedman, *Introduction to Financial Technology.*
4. John H. Allan, "Stock Exchange Gets New Ticker," *New York Times,* October 24, 1963.
5. Ibid.
6. "Exchange to Rush New Ticker System," *New York Times,* August 11, 1929.
7. "High-Speed Stock Tickers to Call for Rise in Rental," *New York Times,* March 2, 1930.

CHAPTER 5: GAMING THE SYSTEM

1. IBM corporate Web site: http://www.research.ibm.com/deepblue/meet/html/d.3.html.
2. Scott Patterson, *The Quants* (New York: Crown, 2010).
3. Sean D. Hamill, "Research on Poker a Good Deal for Airport Security," *Pittsburgh Post-Gazette,* August 2, 2010.
4. Michael Kaplan, "Wall Street Firm Uses Algorithms to Make Sports Betting Like Stock Trading," *Wired,* November 1, 2010.

5. Bueno de Mesquita's Mubarak prediction was fact-checked with multiple sources. The name of the Wall Street firm is not disclosed to honor nondisclosure agreements.
6. IBM chronicles of Deep Blue: http://www.research.ibm.com/deepblue/home/may11/interview_1.html.
7. Tom Pedulla and Rachel Shuster, "Whole World Taken with Knicks' Star Jeremy Lin," *USA Today,* February 11, 2012.
8. Harvey Araton, "Lin Keeps His Cool; Around Him, Heads Spin," *New York Times,* February 11, 2012.
9. Ed Weiland, "NBA Draft Preview 2010: Jeremy Lin, G Harvard," *Hoops Analyst,* May 13, 2010, http://hoopsanalyst.com/blog/?p=487.
10. Nick Paumgarten, "Looking for Someone: Sex, Love, and Loneliness on the Internet," *New Yorker,* July 4, 2011.
11. Ibid.
12. Study by Match.com and Chadwick Martin Bailey, http://cp.match.com/cppp/media/CMB_Study.pdf.
13. John Tierney, "Hitting It Off, Thanks to Algorithms of Love," *New York Times,* January 29, 2008.
14. Andrew Stern, "Researchers Say Dating Websites Make Poor Cupids," Reuters, February 7, 2012.
15. Eli Finkel, Paul Eastwick, Benjamin Karney, and Harry Reis, "Online Dating: A Critical Analysis from the Perspective of Psychological Science," *Journal of Psychological Science in the Public Interest,* February 2012.

CHAPTER 6: PAGING DR. BOT

1. Susan Adams, "Un-Freakonomics," *Forbes,* August 9, 2010.
2. National Kidney and Urologic Diseases Information Clearinghouse (NKUDIC), http://kidney.niddk.nih.gov/KUDiseases/pubs/kustats/index.aspx.
3. Rita Rubin, "Dialysis Treatment in USA: High Costs, High Death Rates," *USA Today,* August 24, 2009.
4. Ezekiel J. Emanuel, "Spending More Doesn't Make Us Healthier," *Opinionator* blog, *New York Times,* October 27, 2011, http://opinionator.blogs.nytimes.com/2011/10/27/spending-more-doesnt-make-us-healthier/.
5. Farhad Manjoo, "Why the Highest-Paid Doctors Are the Most Vulnerable to Automation," *Slate,* September 27, 2011.
6. "BD FocalPoint Slide Profiler vs. Manual Review," http://www.bd.com/tripath/labs/fp_detection_pop.asp.
7. National Cancer Institute at the National Institutes of Health, http://www.cancer.gov/cancertopics/factsheet/detection/Pap-test.
8. S. V. Destounis et al., "Can Computer-Aided Detection with Double Reading of Screening Mammograms Help Decrease the False-Negative Rate? Initial Experience," *Radiology* 232, no. 2 (August 2004): 578–84, http://www.ncbi.nlm.nih.gov/pubmed/15229350.
9. "Screening and Diagnosis," Stanford University Medicine, Cancer Institute, http://cancer.stanford.edu/breastcancer/diagnostic_tests.html.
10. Christina E. Seeley et al., "A Baseline Study of Medication Error Rates at Baylor

University Medical Center in Preparation for Implementation of a Computerized Physician Order Entry System," *Proceedings of Baylor University Medical Center* 17, no. 3 (July 2004): 357–61, http://www.ncbi.nlm.nih.gov/pmc/articles/PMC1200672/.
11. E. A. Flynn et al., "National Observational Study of Prescription Dispensing Accuracy and Safety in 50 Pharmacies," *Journal of the American Pharmacists Association* 43, no. 2 (2003): 191–200.
12. Kevin McCoy, "Lawsuit: Walgreens Prescription Error Killed Man," *USA Today,* November 2, 2007.
13. Vinod Khosla, "Do We Need Doctors or Algorithms?" *TechCrunch,* January 10, 2012, http://techcrunch.com/2012/01/10/doctors-or-algorithms/.
14. Jerome Groopman, *How Doctors Think* (New York: Houghton Mifflin, 2007).
15. Ibid.
16. Fred Herbert, *Looking Back (and Forth): Reflections of an Old-fashioned Doctor* (Macon, GA: Mercer University Press, 2003), p. 37.
17. David Leonhardt, "Making Health Care Better," *New York Times Magazine,* November 3, 2009.
18. Ibid.
19. Ibid.
20. "What Is Heart Failure?" National Heart, Lung, and Blood Institute, http://www.nhlbi.nih.gov/health/health-topics/topics/hf/.
21. "The Power of Knowing," 23andMe, https://www.23andme.com/stories/6/.
22. Andrew Pollack, "DNA Sequencing Caught in Deluge of Data," *New York Times,* November 30, 2011.
23. Ewen Callaway, "Ancient DNA Reveals Secrets of Human History," *Nature,* no. 476 (August 9, 2011): 136–37.
24. Anna Wilde Mathews, "WellPoint's New Hire. What Is Watson?" *Wall Street Journal,* September 12, 2011.

CHAPTER 7: CATEGORIZING HUMANKIND

1. Judy L. Hasday, *The Apollo 13 Mission* (New York: Chelsea House, 2001), p. 16.
2. Andrew Chaikin, *A Man on the Moon: The Voyages of the Apollo Astronauts* (New York: Penguin, 1995).
3. During my first conversation with Capers, which lasted almost two hours, he debriefed me on my own personality at the end of our talk. It turns out that the largest piece of my persona is thoughts-driven. I'm pragmatic and usually make decisions based on facts, but I also have a hard time delegating and can become irate under pressure (the latter part not being a fabulous trait for an author or a tech entrepreneur). When we first started chatting, Capers described how the theory worked, and I followed along, nodding. Truth be told, I wasn't yet sold on this whole thing. How could he know what I was thinking based solely on my words? I could say one thing and mean something completely different.

   In one case, that's what I did—it's something I often do. As Capers continued his early explanations, I often responded with a nod and the reply of "interesting." This utterance, he later told me, clearly signaled that I wasn't impressed and that I was still withholding judgment. "You don't actually mean, 'Hey, that's interesting,'" Capers explained to me. "What you're really saying is, 'Hmm, I'm not sure about that.'"

He was spot on. A thoughts-based person who issues short replies with little emotion is often just trolling for more information. Either that or they're totally uninterested. But Capers said I later perked up when he started telling me about some test statistics. "You said, 'Wow, that's really cool,'" he recounted. "When a thoughts-based person puts an adverb together with an adjective such as 'cool,' it means you're truly impressed."

Indeed I was. Just like everybody else, my personality had mostly been formed by the time I was seven or eight years old.

4. Capers has an interesting story himself, including a career stint in rock and roll. He was one-half of the 1960s–1970s rock group Hedge and Donna, who released six records during a ten-year stretch. His showbiz career also led him to the movies; he starred in 1974's *The Legend of Hillbilly John*. When his Hollywood luck ran out, Capers spent the requisite years in school to become a psychiatrist. He met Kahler in San Diego, where he proved a quick learner of the theory Kahler, McGuire, and NASA had developed.
5. Sebastian Mallaby, *More Money Than God: Hedge Funds and the Making of a New Elite* (New York: Penguin Press, 2010).
6. Peter Brown, Robert Mercer, Stephen Della Pietra, and Vincent J. Della Pietra, "The Mathematics of Statistical Machine Translation: Parameter Estimation," *Journal of Computational Linguistics* 19, no. 2 (1993): 263–311.
7. Ingfei Chen, "Scientist at Work: Nick Patterson," *New York Times*, December 12, 2006.

CHAPTER 8: WALL STREET VERSUS SILICON VALLEY

1. Rana Foroohar, "Wall Street: Aiding the Economic Recovery, or Strangling It?" *Time*, April 4, 2011.
2. While building his team at Teza, Malyshev recruited a Goldman Sachs programmer, Sergey Aleynikov, by offering to triple his $400,000 salary to $1.2 million. Aleynikov agreed to come over and made a hasty move to Chicago. A day after he began working at Teza, Aleynikov was arrested on charges that he had stolen computer code from Goldman. During his last three days of work at Goldman, Aleynikov had systematically uploaded big chunks of Goldman's trading code to a server in Germany. But Goldman's tech security spotted the maneuver and alerted the FBI. In 2011, Aleynikov received an eight-year jail sentence from a federal court. In February 2012, however, he got a break as a U.S. appeals court in New York overturned the conviction after the defense argued that the prosecution had unfairly utilized the Economic Espionage Act in the case.
3. Jacob Bunge and Amy Or, "Ex-Citadel Employee Arrested for Allegedly Stealing Code," Dow Jones Newswires, October 13, 2011.
4. Paul Krugman, "The Market Mystique," *New York Times*, March 26, 2009.
5. The CIA World Factbook, https://www.cia.gov/library/publications/the-world-factbook/.

CHAPTER 9: WALL STREET'S LOSS IS A GAIN FOR THE REST OF US

1. Matthew Humphries, "Facebook Stores up to 800 Pages of Personal Data Per User Account," Geek.com, September 28, 2011, http://www.geek.com/articles/geek-pick/facebook-stores-up-to-800-pages-of-personal-data-per-user-account-20110928/.
2. Ashlee Vance, "This Tech Bubble Is Different," *Bloomberg Businessweek,* April 14, 2011.
3. Peter Cohan, "Head in the Cloud," *Hemispheres,* November 2011.
4. Stu Woo and Raice Shayndi, "EA Invades Zynga's Turf," *Wall Street Journal,* November 2, 2011.
5. Paul Kedrosky and Dane Stangler, *Financialization and Its Entrepreneurial Consequences,* Kauffman Foundation Research Series, March 2011, http://www.kauffman.org/uploadedfiles/financialization_report_3-23-11.pdf.

CHAPTER 10: THE FUTURE BELONGS TO THE ALGORITHMS AND THEIR CREATORS

1. "Algorithm Measures Human Pecking Order," *MIT Technology Review,* December 21, 2011, http://www.technologyreview.com/blog/arxiv/27437/.
2. John Markoff, "Armies of Expensive Lawyers, Replaced by Cheaper Software," *New York Times,* March 4, 2011.
3. Eric Brynjolfsson and Andrew McAfee, *Race against the Machine* (Digital Frontier Press e-book, 2011).
4. Steve Lohr, "More Jobs Predicted for Machines, Not People," *New York Times,* October 23, 2011.
5. Steve Lohr, "In Case You Wondered, a Real Human Wrote This Column," *New York Times,* September 10, 2011.
6. Christopher Drew, "Why Science Majors Change Their Minds (It's Just So Darn Hard)," *New York Times,* November 4, 2011.

INDEX